ANCESTRY OF
Archibald Edward Garrod

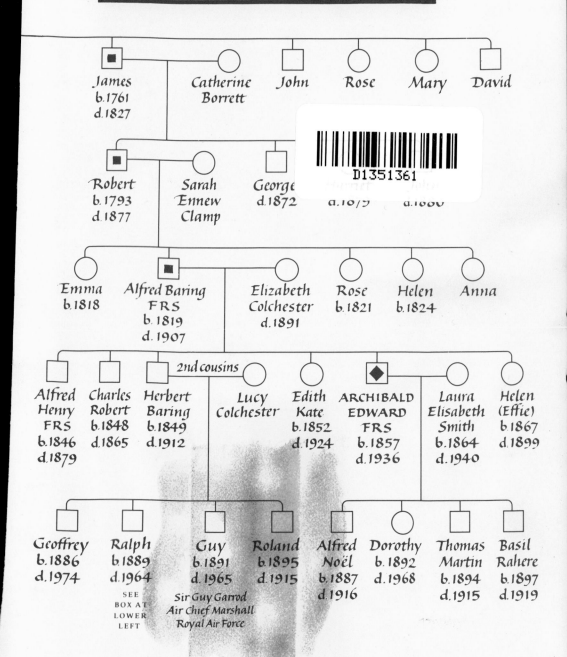

Archibald Garrod
and the Individuality of Man

ARCHIBALD
EDWARD·GARROD

E·H·NEW   A·D·1922

Archibald Edward Garrod's bookplate. Designed by Edmund New, a prominent illustrator of the time, the bookplate depicts Christ Church, Garrod's college at Oxford, and Tom Tower, designed by Sir Christopher Wren and completed in 1682.

Archibald Garrod
and the Individuality of Man

ALEXANDER G. BEARN

The Rockefeller University, New York

CLARENDON PRESS • OXFORD
1993

Oxford University Press, Walton Street, Oxford OX2 6DP

Oxford New York Toronto
Delhi Bombay Calcutta Madras Karachi
Kuala Lumpur Singapore Hong Kong Tokyo
Nairobi Dar es Salaam Cape Town
Melbourne Auckland Madrid
and associated companies in
Berlin Ibadan

Oxford is a trade mark of Oxford University Press

Published in the United States
by Oxford University Press Inc., New York

A catalogue record for this book is available from the British Library

Library of Congress Cataloging in Publication Data
Bearn, Alexander G., 1923–
Archibald Garrod and the individuality of Man/Alexander G. Bearn.
Includes bibliographical references.
1. Garrod, Archibald E. (Archibald Edward), Sir. 1857–1936.
2. Biochemists—England—Biography.   3. Physicians—England—
Biography.   4. Biochemical variation—Research—History.
I. Title.
QP511.8.G37B4   1992   610'.92—dc20   92–27504
ISBN 0 19 2621459

Typeset by
EXPO Holdings, Petaling Jaya, Malaysia
Printed in Great Britain by
Bookcraft (Bath) Ltd, Midsomer Norton, Avon

Dedicated to all those who, like

ARCHIBALD GARROD

have the faith to pursue science in

the quest for a higher medicine.

FOREWORD

Joseph L. Goldstein and Michael S. Brown
Department of Molecular Genetics and Internal Medicine
University of Texas Southwestern Medical Center
Dallas, Texas

As students of biology, we learn that Archibald Garrod advanced the concept of inborn errors of metabolism through brilliant insights into patients with rare genetic diseases. Often we overlook Garrod's two other prescient concepts: (1) the concept of the chemical uniqueness of the individual, which prefigured the modern appreciation of individual predisposition to disease; and (2) the concept of the physician-scientist, which prefigured the modern revolution in biomedical research. All three concepts are brought into sharp focus in this incisive biography, written devotedly by Alexander G. Bearn, a disciple whose own career is based firmly on the Garrodian tradition.

Garrod's insights into disease were made possible by two developments in basic science at the turn of the century: (1) advances in chemistry that allowed him to determine the structures and amounts of organic molecules in the body fluids of patients; and (2) advances in genetics, stemming from the observations of Mendel, that allowed him to interpret the puzzling familial features of the chemical abnormalities that he detected. Garrod used his patients to form a bridge between these two previously disconnected scientific disciplines, thereby shoring up both disciplines and establishing a theme for the physician-scientist as bridge-builder.

Trained as a physician and as a chemist, Garrod had faith only in substances that he could measure physically. Never comfortable with the vague concept of a gene, he nevertheless accepted Bateson's suggestion that the laws that govern garden peas also governed his patients. Thus, a second theme of the physician-scientist was born: an interpreter who can expose the simplicities underlying a complex clinical problem so that a basic scientist can solve it. This theme was repeated in 1945 when William B. Castle, a physician-scientist on the Harvard service at Boston City Hospital, reduced the complex problem of sickle cell anemia to a simple question that could be answered by the chemist Linus Pauling. It has been repeated countless times ever since.

Garrod successfully disseminated his model. At London's St Bartholomew's Hospital, he created a plan for the first unit in England in which a full-time salaried physician-scientist Professor would supervise junior physicians and scientists in the performance of clinical research in laboratories directly connected to in-patient wards and out-patient departments. This entire group would focus on the unknown, as opposed to the known, aspects of disease. This model was subsequently adopted in many American medical schools with generous financial support from the National Institutes of Health, and it has led to innumerable medical triumphs.

The foundations for bridge-building in medicine have never been more solid than they are today. Like the chemists of Garrod's day, modern molecular biologists have created new tools that allow physician-scientists to apply exciting new science to every human disease. Yet paradoxically, just when opportunities most abound, the Garrodian model is eroding. Clinical units, faced with increasing pressure to obtain income from patient service, are focusing more on the known, as opposed to the unknown, aspects of disease. At the same time, young physician-scientists are lured away from clinical medicine by the quick rewards of glamorous basic science experiments, which are technically simple to perform owing to the commercial availability of reagents and equipment. These experiments do not require the simplifying insight that is essential for pioneering clinically-based research. Garrod foresaw this erosion in 1919 in a remarkably prescient lecture on the occasion of Sir William Osler's 70th birthday. He said:

... there is still much work ahead for clinical physicians and surgeons, in the advancement of knowledge as well as in the treatment of the sick, and especially for concerted work in which laboratory and ward workers co-operate as colleagues, and without any claim to a monopoly of the scientific spirit and method on either side....

If it were possible to concentrate the whole of the scientific work of medicine in the laboratories, and if clinical studies came to be regarded as unattractive to men of scientific instincts, the results would be deplorable for medical practitioners and patients alike. Even now there are signs of diminished zeal on the part of students to become adept at the purely bedside methods of examination. Unless the whole field of medicine be permeated by the scientific spirit we can look for little progress, and shall return to the conditions in the Middle Ages, in which practitioners merely accepted and applied what was to be learned by study of the writings of Hippocrates and Galen.

Hopefully, this book will help to prevent erosion of the Garrodian model by offering us an understanding of the qualities that made Garrod a successful physician-scientist. First, Garrod was a serious physician whose intellectual stimulation came from individual patients and whose creative insights emerged from a thorough understanding of their illnesses. Second, he made clinical observations with sufficient power and precision to be understood by basic scientists such as Bateson, the geneticist, and Hopkins, the biochemist. Third, Garrod resisted the temptation to be drawn away from patients and into non-clinical research, even though his research skills would clearly have permitted such a diversion.

Why is the Garrodian model failing? Largely because it is difficult to find and support physician-scientists who have the right balance to lead clinical departments in medical schools. At one extreme, we have pure clinicians who lack the scientific training necessary to formulate problems in ways that direct the focus of basic scientists. At the other extreme, we have scientists who lack the clinical training and experience necessary to expose the core of a disease. Lacking the leadership of scientifically erudite clinicians, young physician-scientists join the race to answer the next obvious question in some fast-moving field of basic science, abandoning efforts to solve any relevant clinical problem.

The publication of this biography, coupled with the recent reissue of Garrod's monograph *The inborn factors in disease* (with its insightful comments by Charles Scriver, Barton Childs, and Joshua Lederberg)*, should give students a glimpse into the challenges and rewards of life as a physician-scientist. By paying tribute to the past, we open possibilities for the future.

*Scriver, C. R. and Childs, B.(1989). *Garrod's The inborn factors in disease.* Oxford University Press, Oxford.

PREFACE

DARWIN'S great work, *On the origin of species*, was published in 1859, and nine years later his second great book, *The variation of animals and plants under domestication*, appeared. For thirty years thereafter, little was done to pursue the genetic clues that lurked in these publications, at least in part because of what the botanist William Bateson called the 'apathy characteristic of faith'.

This is not to say that nothing was learned during those years. Between 1870 and 1900 chromosomes became established as the root of heredity, and many theories were advanced to explain how certain characters were passed from parents to offspring. It became known that every cell nucleus contained the same number of chromosomes; that the sperm and the egg contributed equal numbers of chromosomes during fertilization; and that the chromosomes as well as the nucleus divided during mitosis, producing two nuclei, each with the same number of chromosomes. It was recognized that when reproductive cells made their final two divisions, each egg and sperm had only half the full number of chromosomes to contribute to the gamete.

In 1900, some forty years after its appearance in an obscure Austrian journal and sixteen after Mendel's death, Mendel's work was 'rediscovered', and scientists began to view their work from another perspective.

In that year, a young and aspiring London physician named Archibald Edward Garrod, who in his spare time was investigating the chemical alterations that occur in disease, turned to an examination of the pigments present in normal and pathological urine. In May 1899 he had read a paper to The Royal Medical and Chirurgical Society of England entitled 'A contribution to the study of alkaptonuria,' a condition characterized by the excretion of 'black urine' from the time of birth.

Although Garrod's paper was principally devoted to a discussion of the chemical abnormalities found in the urine in this rare condition, he noted that in many of his patients, as well as those described in the literature, more than one family member was affected by the disease.

Garrod became intrigued by the possibility that alkaptonuria might be inherited, but was unable to find an instance where the condition was transmitted from one generation to the next, as had been shown to occur in a number of other inherited disorders, such as brachydactyly.

Garrod was soon to strike up a friendship with William Bateson, an early and enthusiastic disciple of Mendel, who quickly resolved the problem. Bateson suggested that the increased consanguinity among the parents of Garrod's patients, as well as the increased frequency of the condition in sibs, could be readily accounted for by assuming that alkaptonuria was inherited in an autosomal recessive fashion. Stimulated by this initial observation, Garrod discovered additional inborn errors of metabolism, and these formed the basis of his now famous Croonian Lectures on *Inborn errors of metabolism*, delivered to the Royal College of Physicians in 1908 and published in 1909.

Inborn errors of metabolism are now a medical commonplace, and Garrod is given ample credit for their initial discovery. Probably his greatest insight, however, was his appreciation of the importance of bio-chemical individuality — an individuality that may be disclosed by an idiosyncratic reaction to a foreign antigen or a drug, or, more generally, by susceptibility or resistance to a variety of human diseases. In his second book, *The inborn factors in disease*, published in 1931, Garrod elaborated on this theme, and articulated the majestic concept of human biochemical individuality.

As we approach the twenty-first century, physicians and other biological scientists widely acknowledge the central importance of Garrod's concept of inborn errors of metabolism. In contrast, his role in formulating the doctrine of biochemical individuality, as well as his contention that human evolution necessarily depends on the existence of biochemical polymorphism, is seldom appreciated.

Garrod's contemporaries in the medical profession widely honored him for his academic success. They respected his medical scholarship, while tacitly acknowledging that he was not an outstanding clinician. Few who attended his named lectures and addresses, or who knew him as a professional colleague, 'heard' the lesson of individuality he loved to preach: disease can only be properly studied in the light of an individual's genetic susceptibility, and that in turn rests on biochemical individuality. Why is it that some individuals are able to adapt successfully to environmental perturbations, while others are not? Not to reflect on such matters, Garrod believed, was to miss an essential biological truth.

It was Archibald Garrod who saw that it was artificial and counter-productive to treat genetics, biochemistry, and medicine as discrete, non-interacting disciplines. It was he who penetrated the vague concepts of idiosyncracy and constitutionality to find their essential biological meaning. Only by thinking of human diseases as the consequence of genetic and environmental interaction was a new horizon in biological science and medical practice made possible.

Today Garrod's position as the father of biochemical genetics is secure. Although the range of human biochemical individuality knows no end, it surely had its beginning in his creative mind. Garrod was not always right, but his contributions were seminal: there is no need to clothe them now with mythic insights. Archibald Garrod is an enduring symbol of a clinically trained physician who dared to look ahead, and who believed, long ago, that clinical medicine must be rooted in science and the scientific method.

New York   A.G.B.
November 1992

ACKNOWLEDGEMENTS

IN 1922, Archibald Garrod delivered the Lettsomian Lectures to the Medical Society of London and chose as his subject glycosuria. When I was invited to give the same Lectures more than fifty years later, I borrowed for my title Garrod's immortal phrase 'Inborn errors of metabolism'. During the preparation of those lectures, I became attracted to the idea of writing a short scientific biography of Garrod, a man whose ideas were relatively unappreciated by his contemporaries and largely ignored by succeeding generations. Only in recent years have I at last found the time to give this project the attention it deserves.

The philosopher Stuart Hampshire has cogently observed that one cannot truly know a person and his inner thoughts until one has seen his face, the look in his eyes, the way he walks and stands, and the varying tone of his voice. Deprived of this opportunity, I have tried to recapture the essence of Garrod's personality and scientific contributions by making extensive use of his own words, his letters, the articles he wrote, and the opinions of his contemporaries. I have tried to avoid ascribing to Garrod thoughts and ideas which are but projections of contemporary knowledge and I have endeavored to put his life and scientific contributions in historical perspective, but I am not a medical historian and must beg the indulgence of those who are.

In writing this book, I have greatly benefited from many conversations with friends and colleagues, and I thank them not only for their wise advice, but for their generosity in the face of my relentless and unceasing enquiry.

Among those to whom I am especially indebted for help, I particularly wish to mention Sir Richard Bayliss for matters relating to the College Club of the Royal College of Physicians; Daniel J. Kevles and Robert E. Kohler for their scholarly insights into the evolution of genetics and biochemistry in Europe and the United States; Robert Olby for his insightful writing about modern development of genetics; Norman Horowitz for insights on the one gene–one enzyme hypothesis; Shigeru Sassa for sharing with me his encyclopedic knowledge of porphyrin metabolism; the late Sewall Wright for his warmth and hospitality on my several trips to Madison, Wisconsin, to obtain his perspective on the emergence of biochemical genetics in the United States; and Tibor Vasko for his knowledge of Czech and his enquiries at the Charles University, Prague. Like so many others concerned with the history of biochemistry, I owe a particular debt of thanks to Joseph Fruton and his scholarly writings on the evolution of German science in the nineteenth century. Victor A. McKusick deserves my thanks for his unstoppable and valuable compilations and catalogues of genetic traits. I also wish to thank Jan Sapp for reading and commenting on an early draft of the manuscript, and John H. Edwards for conversations not always understood but invariably stimulating.

I am grateful to the late Sir Charles Symonds for generously providing information on Garrod's years in Malta, Madeleine Lovedy-Smith for her reminiscences about the family particularly Dorothy Garrod, R. Macbeth for lively recollections of Garrod as Regius Professor of Medicine, Lady Peters and the late Sir Rudolph Peters for their retrospections on the relationship between Garrod and the development of the Department of Biochemistry at Oxford, Jacquetta Hawkes (Mrs J. B. Priestley) for her reflections on Archibald Garrod toward the end of his life, and Carl Birger van der Hagen for his knowledge of the fjords of Norway.

Frank Vella generously made available to me detailed information regarding Garrod's time in Malta. His help has been invaluable. Alastair Robb-Smith's authentic knowledge of Oxford medicine was of critical value to me in writing the chapter dealing with Garrod's years at the University.

Professor Henry Harris, Regius Professor of Medicine at Oxford (1979–1992), in addition to providing me with most useful information on the nature of the Regius Chair, encouraged me to undertake the writing of this book. I also wish to thank Helen J. Jordan for her role in initiating this project and Charles R. Scriver for his continued encouragement.

I am much indebted to my late brother-in-law, Leonard Porter, for his meticulous and indefatigable historical research on the ancestry of Archibald Garrod. The fruits of his considerable labors appear as an appendix. I deeply regret that he did not live to see the completion of this book.

Professor A. Norman Jeffares provided sage advice, and his rollicking good humor forever buoyed me up. Norton Zinder's cafeteria conversations have been always enlightening. Jesse Ausubel's constant interest has been constructive and cheering.

Joshua Lederberg has enthusiastically supported me at every turn. In addition to providing me with a tranquil office overlooking New York

City's East River, he has been of particular help in sharing his insights into the early days of bacterial and *Neurospora* genetics.

The hospitality and helpfulness of archivists and librarians have been of inestimable value, and I wish particularly to thank John Ayres, Deputy Librarian of the Royal Society of Medicine; Nicholas Baldwin and Eileen Power, Archivists for the Hospital for Sick Children at Great Ormond Street; Brian J. Harrison and Rosemary Harvey and the archivists at the John Innes Institute, Norwich; Sonya W. Mirsky, Patricia Mackey, and John Wilson, at the Rockefeller University Library; Darwin H. Stapleton and Thomas Rosenbaum of the Rockefeller Archive Center; Sheila Edwards and her colleagues at The Royal Society; Geoffrey Davenport, Librarian, Royal College of Physicians; Eric J. Freeman, Librarian, and Julia Sheppard, Archivist, The Wellcome Institute for the History of Medicine; Judith M. Etherton, St Bartholomew's Hospital; June Wells, Christ Church, Oxford; David R. C. West, Archivist, Marlborough College; the Bodleian Librarian, Oxford, and the Curators of the Library, as well as Simon Bailey, the University Archivist.

For skillful and painstaking editorial assistance I am deeply grateful to Blair Potter for many substantial and clarifying improvements in the manuscript. Reynard Biemiller's talented and artistic hand is responsible for the pedigree displayed as the end papers. I am also indebted to Helene Friedman, who enabled me to meet the deadline, and to Meta Wyss and Cathy E. Harbert for library support. Kathleen Chopin adroitly collated the illustrations, and I am grateful for her perceptive eye. Gordon Bearn has tried, without always achieving complete success, to instruct me on matters philosophical.

It is a particular pleasure for me to thank my long-time friend Barton Childs who, in addition to a most careful and critical reading of the entire manuscript, encouraged me during periods of glacial progress. I have greatly profited from our many stimulating conversations and his thoughtful and constructive comments.

My greatest debt is due to the late Dr Oliver Garrod and to his son, Professor Simon Garrod, who not only made available unpublished letters and various memorabilia of Archibald Garrod but also extended to me every conceivable assistance over a period of several years. Without their patience and generosity, this book would never have been written. Sir Reginald and Lady Murley also kindly shared their knowledge of the family.

In 1949 while still a medical student at Bart's, Dr C. J. R. Hart was awarded the Wix Prize for his scholarly essay on Garrod. He has generously allowed me to use material he collected at that time.

The generous support of the Alfred P. Sloan Foundation, the Richard Lounsbery Foundation, the Archbold Charitable Trust, and the Commonwealth Fund was of critical importance to the completion of this work. My administrative assistant, Jan Maier, has uncomplainingly typed innumerable drafts and has been tenacious and resourceful in tracking down elusive references. I am most grateful for her talented and tireless support.

I am deeply grateful to Joseph L. Goldstein and Michael S. Brown for generously agreeing to write the Foreword for this book. I am also delighted that they share my admiration for Archibald Garrod's insight, wisdom, and scientific perspicacity.

My wife, Margaret, has prodded me on in the nicest possible way. She has cheerfully insulated me from innumerable domestic chores as well as providing intellectual encouragement and creative criticism. Her companionship and perpetual support made the writing of this book not only possible but enjoyable.

CONTENTS

1

The Family

THE London that welcomed Archibald Edward Garrod on 25 November 1857 was a city of strident contrasts. It was the greatest urban center in the world, the bustling, prosperous heart of the vast British Empire. Warehouses, railroads, and docks serviced the constant traffic on the River Thames. Yet, in this city of three million, economic prosperity rubbed shoulders with incredible squalor. The crime rate was inordinately high. Prostitution was rampant — even Parisians branded London 'the City of Sin'. Open sewers in the slums encouraged cholera, which had spread to England from India: in the epidemic of 1849, some 13 000 persons had died of the disease in London alone. Joseph Lister was soon to wage an uphill fight for antiseptic treatment of gangrene, erysipelas, and other infections that were the scourge of hospitals.[1] Although Antoni van Leeuwenhoek had seen his 'very little animalcules, very prettily moving' 175 years earlier, their relationship to disease had been but little considered.[2]

In mid-century, London boasted 12 charitable hospitals with a total of nearly 4000 beds; these and other institutions treated about half a million out-patients yearly. At best this treatment was uncertain; most often it was ineffectual. There were virtually no trained nurses. In 1854, at Scutari in the Crimea, Florence Nightingale had just begun her own battle for proper nursing care; there, in the first long winter of the Crimean War, two of every five wounded soldiers died. In London, many of the nurses were prostitutes, alcoholics, criminals, or illiterates — as was Charles Dickens's Sairy Gamp, one of many creations born of his keen awareness of the need for social reform. In 1860, with characteristic zeal, Florence Nightingale established the first training school at St Thomas's Hospital, to 'recruit women of character and intelligence' into nursing.[3]

We do not know how much the seamy side of the city impinged on the Garrods at 84 Harley Street. Certainly the miserable health of much of the population must have been abundantly apparent to Archibald's father, Alfred Baring Garrod, an eminent physician who viewed many of the city's horrors daily in the course of his hospital duties.[4]

Archibald was born into a remarkable Suffolk family of rising social stature. His grandfather, Robert Garrod (Fig. 1), the son of a tenant farmer in Stradbroke, was a man of exceptional ability and drive.[5] Apprenticed to his father's landlord, a brewery owner in Ipswich, Robert later founded a successful firm of auctioneers and estate agents, which continued to exist as a separate business until it was

Fig. 1 Robert Garrod, Archibald's grandfather, *circa* 1820.

absorbed a few years ago. Alfred Baring Garrod (Figs 2 and 3) was born in Ipswich, as was Archibald's mother, Elizabeth Ann Colchester. Alfred attended the local grammar school; but despite his father's thriving business he decided early in his boyhood that he wanted to become a doctor.

In England at that time medicine was not highly regarded as a profession; it certainly did not enjoy the social status of the army, the law, or the church. Those careers were largely reserved for young men entitled to be called 'gentlemen'. Nevertheless, medicine was beginning to lose the stigma of trade, and it was a good way for ambitious tradesmen and intelligent artisans to improve their social status. Alfred

Fig. 2 Sir Alfred Baring Garrod, Archibald's father, *circa* 1885.

Fig. 3 In 1899 the French town of Aix-les-Bains named this street for Sir Alfred Garrod. In their dedication, the municipal council honored his writings on gout, which stressed the value of the waters at Aix, and brought at least 1200 new patients to take the treatment there. Indeed, the council felt that Garrod was responsible for making a sojourn at Aix standard in the treatment of gout.

embarked on his career by becoming apprenticed to his father's neighbor and political associate Charles Chambers Hammond, a Suffolk physician, and continued with formal medical education at University College Hospital at the University of London, where he received his MD in 1843.

Alfred was exceptionally talented, and soon established himself as an able physician and scientist. In 1848, at the age of twenty-nine, he made a seminal discovery for which he is still well known. In a paper delivered before the Royal Medical and Chirurgical Society, he described the presence of uric acid in the blood of his patients with gout.[6] With considerable skill, he developed a simple chemical test that could be conveniently undertaken using a relatively small sample of blood. In addition, he demonstrated that patients with so-called rheumatic gout, although they were quite as incapacitated as patients with classical gout, had no increase of uric acid in their blood. With this discovery, he made the first distinction between gout and rheumatic arthritis — a term he introduced to replace the misleading 'rheumatic gout'.

Alfred Garrod's interest in rheumatic diseases continued almost uninterruptedly throughout his professional life. His book *The essentials of materia medica, therapeutics, and the pharmacopoeia*, published first in 1855, went through 13 editions, and earned him an enviable reputation in rheumatic diseases.[7] Four years later he published another classic, *The nature and treatment of gout and rheumatic gout*.[8] Distinctions and honors were showered upon him, and he was invited to deliver the most prestigious lectures of his day. In 1858, at the early age of 39, he was elected a Fellow of the Royal Society. Two years later, in 1860, Joseph Lister and Garrod's friend and neighbor Francis Galton were also to be elected Fellows.

Alfred Garrod left University College in 1862, after nineteen years on the teaching staff, and was appointed physician to King's College Hospital in South London and professor at King's College in the Strand. In 1874, at the age of fifty-five, he gave up hospital work, but continued his lucrative consulting practice, as well as his work on the role of chemistry in the elucidation of disease. He received a knighthood in 1887, on the occasion of Queen Victoria's first Jubilee, and shortly thereafter was appointed one of the Physicians Extraordinary to the Queen. An increasingly wealthy man, he died on 28 December 1907 in his eighty-ninth year, leaving an estate of £84 551. In addition to what he left his family, Sir Alfred left legacies to his coachman, butler, and cook, as well as to his other servants who had been with the household for five years or more.

Sir Alfred Garrod's extraordinary intellectual commitment to the advancement of chemical pathology undoubtedly influenced his son Archibald, who, on qualifying in medicine, pursued his father's interest in rheumatoid arthritis. Indeed, in 1890, when Archibald Garrod was only thirty-three years old, he published a work of significant scholarship entitled *A treatise on rheumatism and rheumatoid arthritis*, which must have afforded great delight to his elderly father.[9]

Alfred and Elizabeth Garrod had five other children: three boys and two girls. Charles Robert, the eldest, was born in 1845 and died of tuberculosis at the age of seventeen. Alfred Henry, the second child, was born in 1846, and went on to achieve a career of exceptional brilliance.[10] He was educated at University College School and in October 1862 entered University College, London, where he was much influenced by Professor William Sharpey's lectures in physiology. In 1868 he graduated in medicine from King's College, London. The practice of medicine, however, held little fascination for Alfred Henry, and, having also won that year an exhibition in natural science at St John's College, Cambridge, he decided to pursue a career in science. In 1871 he was placed senior in the Natural Science tripos, the final honours examination for the BA degree at Cambridge.[11]

Two years later Alfred Henry was elected to a fellowship at St John's — the first time this distinction was ever given to a student in natural science. Throughout his undergraduate years he undertook research on the circulation of the blood. During this time he also made substantial improvements to the sphygmograph, a newly devised instrument that, when applied over a peripheral artery, recorded the blood pressure. In addition, he also recognized that alteration in the temperature of the body was influenced by the dilation and constriction of arteries and other blood vessels.

In 1872 Alfred Henry was elected prosector to the Zoological Gardens in London, where he remained for the rest of his academic career. Today, he is remembered chiefly for his reclassification of several species of birds, based on meticulous anatomical studies. He also made far-reaching contributions to the classification of ruminant mammals. Holding some of the most important academic posts in his field, Alfred Henry served as the Fullerian Professor of Physiology at the Royal Institution and Professor of Comparative Anatomy at King's College. He wore his scholarly learning and intellectual brilliance lightly. Widely acknowledged as a colorful and popular lecturer, he was also a subeditor of *Nature*, and was elected a member of the Royal Society at the age of thirty, the first year that he was proposed (Fig. 4).

In June 1878, with no prior warning, and with his career in full flood and a future full of promise, Alfred Henry suffered a severe pulmonary hemorrhage caused by tuberculosis. He ignored this danger sign, and continued to work unceasingly. It was only with considerable reluctance that he agreed, in December 1878, to enter a sanatorium in Switzerland in an attempt to arrest the disease. Impatient at seeing no improvement, he insisted on returning home to resume his studies. Although he continued to work almost until the end of his life, his condition deteriorated rapidly. He died on 17 October 1879, at the age of thirty-three, in his parents' home.

Although Alfred Henry was ten years older than Archibald, he had a considerable influence on his brother's intellectual development and scientific interest. Alfred Henry did not marry, and the two lived with their parents, at their new house at 10 Harley Street, until his death.

Fig. 4 Alfred Henry Garrod, Archibald's brother, in 1876, the year he was elected a Fellow of the Royal Society (National Portrait Gallery, London).

The Garrods' third son, Herbert Baring, was eight years older than Archibald; and he, too, was unusually able.[12] A classical scholar at school and college, he won prizes for Greek, Latin, German, French, and Hebrew. The ease with which he accumulated prizes earned him the nickname 'the cormorant', but the work came easily to him; he was never one to strive for prizes for their own sake. He left school to take up a scholarship at Merton College, Oxford, where he won one of the university's most prestigious prizes, the Newdigate Prize for English verse, with a poem on Charlemagne.

Herbert's scholarship was of a high order. He achieved an international reputation as a classical scholar, and was widely regarded as an authority on Dante, Goethe, and Calderón. He enjoyed writing for a general audience as well as for his fellow scholars, and was a frequent contributor to *The Spectator*, which was edited by his uncle Meredith White Townsend.

Herbert married his second cousin Lucy Florence Colchester, who added her own spirited and intellectual accomplishments to this enormously gifted family. Lucy was an ardent suffragette, and marched arm-in-arm with her friend Emmeline Pankhurst, the indefatigable and militant champion of women's suffrage whose battles landed her in jails off and on for fifty years. After Herbert's death in 1912 at the age of sixty-three, Lucy traveled widely on the Continent and in North Africa, South America, and the Far East — distinctly avant-garde behavior for a woman of that period. For a while she lived in Korea, and then spent a year in Japan, where she was the honored guest of the Chief Magistrate of Tokyo, who had, in his youth, been one of Herbert's students.[13]

Archibald had two sisters, Edith Kate and Helen Effie, neither of whom married. Effie was the youngest of the family. She lived at home, and died of disseminated tuberculosis in 1899, at the age of thirty-two. Edith also lived at home and kept house for her father after her mother died. On her father's death in 1907 she moved to Ockham, Surrey, to live in the countryside; there, surrounded by books, she lived the life of an independently wealthy, civic-minded countrywoman. Edith died of pancreatic carcinoma in 1924 at the age of seventy-two. She had always been close to her brother Archibald; and in her will she bequeathed money for the foundation of a medical scholarship to the London School for Medicine for Women.

Archibald's childhood coincided with the zenith of what has been called 'The English Century'. In 1857, the year of his birth, Queen Victoria opened the Science Museum at South Kensington, and *Little Dorrit* was being serialized in the monthly magazine *Household words*, adding to Charles Dickens's already overwhelming popularity.

The natural sciences were in ferment as never before. From correspondence with his friend Alfred Russel Wallace, Charles Darwin was learning that their ideas were similar: 'I can plainly see that we have thought much alike and have come to similar conclusions.'[14] Although Darwin's *On the origin of species* would not be published until 1859, he was already putting pen to paper in the year of Archibald's birth. In the field of organic chemistry, as Joseph Fruton[15] has pointed out, new ideas about the structure and function of organic molecules were emerging. A rapidly growing field, organic chemistry would later capture Archibald's interest and lead him to study human disease with the tools of chemistry.

Archibald's parents ran a quintessential upper-middle-class Victorian academic household. The atmosphere was predominantly intellectual, and the children, particularly the boys, were expected to pursue scholarly careers. According to George Graham, a friend and junior colleague, Archibald disliked his given name intensely. It had been selected by his mother, who had read in the newspapers that Archibald Campbell Tait, Bishop of London (later Archbishop of Canterbury) and one of the foremost public men of his day, had lost five children to virulent scarlet fever within five weeks.[16,17] Nine months later, Elizabeth Garrod was still so suffused with pity for the bishop that she decided to name her son Archibald in remembrance of his loss. Elizabeth always called her son by his nickname Archie, as did the rest of his family and their close friends. He received a broad

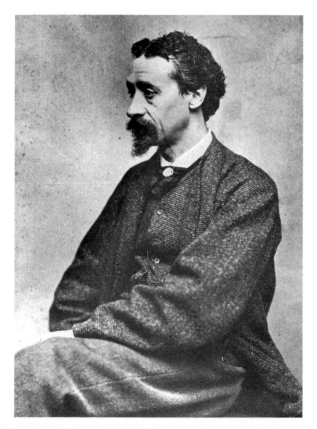

Fig. 5 Charles Samuel Keene, Archibald's second cousin and a frequent visitor to the Garrod home. One of England's great draftsmen. Keene's cartoons often appeared in *Punch*.

Fig. 6 Sir Francis Galton, another familiar face in the Garrod household. Galton's *Hereditary genius*, published in 1869, was much admired by Charles Darwin. Galton is recognized as a pioneer in the study of heredity.

education at home, not only from the members of his own family, but also from their many friends.

Among the distinguished persons of the day who dined with the Garrods at Harley Street were Archibald's second cousin Charles Samuel Keene[18] and his uncle Meredith White Townsend.[19] Keene, a handsome man with a Shakespearean beard and dark eyes (Fig. 5), was a close friend of Archibald's brother Alfred, and a frequent visitor. One of England's greatest draftsmen, Keene worked in pen and ink, in watercolor, and with needle on copperplate. His first published illustration appeared in an edition of *Robinson Crusoe*, and his drawings were regular features in a number of illustrated periodicals. He illustrated Charles Reade's *The cloister and the hearth* and George Meredith's *Evan Harrington*. But Keene's popularity rested primarily on the cartoons he contributed to *Punch* for almost forty years. Humorous and ironic, his exaggerated and genial portrayals of waiters, gamekeepers, tradesmen, and bus-drivers were enjoyed by a wide readership. Keene often arrived at Harley Street with his satchel, carrying an original

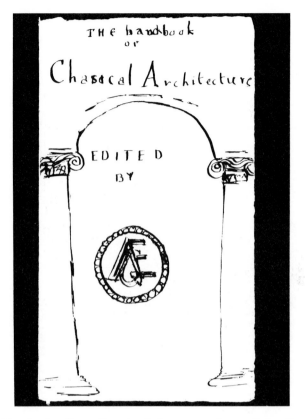

Fig. 7 Within the intellectually stimulating atmosphere of the Garrod home, Archibald's first scholarly efforts appeared early. He produced the meticulously researched and ambitiously illustrated *Handbook of clasical* [sic] *architecture* at the age of ten.

cartoon from *Punch* as a gift for his cousins. It was largely through Keene that Archibald became exposed to the world of illustration and art at an early age.

A less frequent visitor was Sir Francis Galton, a contemporary of Archibald's father (Fig. 6). Galton, a cousin of Charles Darwin, was a distinguished scientist and a pioneer in the study of heredity. In his classic *Hereditary genius*, published in 1869, Galton analyzed men from famous families, and emphasized — often overemphasized — the role of heredity in their development.[20] His book was a provocative account of biological determinism which Charles Darwin greatly admired.

Fig. 8 In the Victorian tradition, young boys and girls alike were intricately coiffed and daintily frocked. Here is Archibald at the age of two, all dressed up for his portrait.

The brilliant household did not awe young Archie; indeed, his family, particularly his brothers, challenged and encouraged him. At the age of ten he began to emulate Charles Keene by writing a rather studiedly solemn illustrated booklet entitled *The handbook of clasical* [sic] *architecture*[21] (Fig. 7). The preface begins, 'This book is intended

Fig. 9 Archibald at the age of five. While slightly more distinguishable, his gender is still largely concealed, in the style of the period.

for the distinction of Grecian architecture from the other kinds.' Archie goes on to document the precise architectural differences among Doric, Corinthian, and Ionic columns. He illustrated the booklet admirably, and recorded some rather lofty personal observations: 'The largest fluted corinthian column that I know is the Nelson column in the Trafalgar Square. The corinthian is fluted like the

Fig. 10 At the ripe age of thirteen Archibald is finally attired in more manly garb.

Ionic.' Instead of signing the work with his full name, he used a signet motif incorporating his initials, AEG. This charming literary excursion must have endeared Archie to his parents and given them hope that he, too, would continue the family tradition of scholarly endeavor. The boy was certainly off to a good start. (Figs 8, 9, and 10.)

2

◆

Growing Up

IN the tradition of most upper-middle-class English families, the Garrods sent Archibald to board at a preparatory school in Harrow, Middlesex, just outside London. But, unlike many youngsters away from home, Archie seemed to enjoy himself at his prep school, and soon developed a keen interest in natural history. In May 1868 he wrote two letters to his mother. Although, as the letters show, his spelling and punctuation at the age of ten are far from perfect, his enthusiasm is beyond doubt:

5.20.68

Dear Mamma

I am very happy and am enjoying myself very much. The other day I found a privet moth crisilis under a privet bush in Mrs wilsons garden the boy had dug it up 4 days before and though it was a dead catapillar. Daniel has just give me another catapillar of the tiger moth. it is a much better specimen than the last we have (alfred & I) found a lot of oak egar catapillars they are gold and black many thanks for the butterfly boxes and the setting board. did Herbert receive my letter I suppose not as I have not had an answer if not please tell me and I will write a letter to the same end but if he did get it ask him to write by the next post if he can today I bough a butterfly net that is mettle soldered on a piece of wood with tin at the top. Aunt Emma is going to make a net for it of green leno Alfred and I are to the field at dale hall tomorrow to catch butterflies he with his net and I with mine. please write by the next post and send me some stamps. the fish now believe me are all quite well now believe me your affectionate son

Archibald Edward Garrod

PS give my love to baby Alfie Herbert Edith and yourself not forgetting papa and 6 kisses for baby x x x x x x.[1] (Fig. 11)

A week later he wrote again:

Dear mamma

I am writing in the end part of a dreadful thunderstorm. Mrs Hollick's farm caught fire at the first flash in about the middle were two awful flashes the thunder of which quite shook the house. The poor servant were in a dreadful state and hid themselves behind the napkin press so that they could not see the lightening the storm

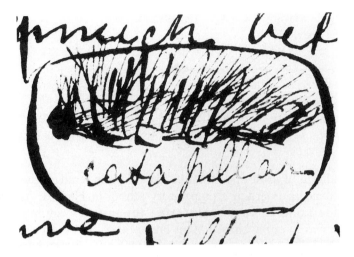

Fig. 11. An illustration in a letter from Archibald, aged ten, to his mother, in which he vividly describes numerous caterpillar sightings. His skills of scientific observation are already becoming evident.

contained both faulk and sheaat lighening there is quite a pond round the middle plot of grass. This morning till twelve o'clock it was quite fine I went out butterfly catching and caught 1 orange tip (male) 1 female 1 male gate keeper 1 small heath one wing of which is shorter than the others. On Wednesday I went to the dale hall fields and caught 2 common blue 3 small heath butterflies I caught a great many more heaths if I had loked.

I enclose a picture of the male and female orange tip both from nature. There is a great deal of differance between the male and female the rain is making an awful noise now. Tell Herbert that there is to be a boat race tomorrow in which Valentine Colchester [a cousin] is going to row there will be three boats each containing five men the prize is a two ginea cup to every man on the first boat. A flash of most vivid blue lightening has just come in at the door accompanied by a peal of thunder. Aunt Emma's much obliged for the coat and trousers. I must now say good by beleive me your affectionate son

Archibald Edward Garrod
another flash and thunder[2]

By the age of twelve Archie had become a tireless collector of butterflies. He described the various species he encountered in the fields and meadows in meticulous detail, and eventually built a very substantial collection, which he donated to Marlborough College. During his Easter holiday at the seaside town of Brighton he wrote home with a bubbling excitement, for he had seen and collected many different butterflies, '… amongst which was a splendid specimen of the orange tip and a green-veined white and I very nearly caught a holly blue, all of which you will see with the names under them in my cabinet.' In a more quantitative and instructional vein he went on, 'You catch one female to 20 males which proves it is much rarer than the males.'[3]

At about the same time that Archie was puzzling over the sex ratio of butterflies, he wrote his second booklet, *The tiger* (Fig. 12). Like his first booklet, *The tiger* is illustrated, but its deliberate construction seems to foretell future scholarship. *The tiger* is divided into four chapters, complete with titles, subtitles, and three footnotes, the last of which points out that Baccus [*sic*] is 'The wine god'. Archie was subject to the normal lip-smacking bloodthirstiness (as well as the still shaky punctuation, spelling, and grammar) of most small boys:

When a tiger has taken an animal like an ox he generally drags it along the ground for a good way. Then eats some after which he lies down and goes to sleep, when he wakes he finishes devouring his prey the husbandman who the ox belonged to then finds the carcass and waits till the tiger is gone away to sleep, when he makes a wound in the half eaten animal and puts in some poison when the tiger eats this his doom is sealed when the poison begins to act he lies down and dies kicking and strugling.

BENGAL TIGER

Fig. 12. Archibald included this particularly accomplished drawing in his booklet *The tiger*.

He charmingly intertwined observation and imagination:

A tiger once appeared at a public meeting of ladies in India. Everyone was frightened but one lady, however, had the presence of mind to face the tiger and kept opening and shutting her parasol in his face: the tiger soon disappeared into a neighbouring jungle.

In one section, he seemed intrigued by the transmission of inherited traits:

There has been an instance of a lion being the father and a tigeress the mother of cubs the lion was born in captivity and the tiger was captured at a very early age the cubs had the head of the lion but the tigerine stripes on the body. The lion–tigers before mentioned are not the only animals of that kind that have ever existed for there have been many examples of this union between the lion and tiger and the mixture of the two animals in their offspring.[4]

Young Garrod was clearly set on a scientific course.

Although Archie enjoyed the out-of-doors and all aspects of natural history, he did not do well in classics. This disquieting news reached his brother Alfred, who wrote to him from St John's College, Cambridge, on 14 November 1868, a few days before Archie's eleventh birthday:

I hear that you do not love those books on the Latin and Greek languages, which you have to work at so much as most boys of your age. There is a man who has been up here 5–6 years and has not passed his little go [entrance examination to Cambridge] yet. He went to a tutor and, when the latter remarked that he could never translate well, unless he knew the grammar first, when he said that he considered that to love the dead languages the only thing necessary was to enter into the spirit of the authors, but he has not passed yet.[5]

Alfred was fond of his little brother and, perhaps sensing that he had been a little hard on him, ended his letter with a conundrum:

I heard a conundrum which perhaps will amuse you. How can you make an egg angry? The answer to the riddle: By calling them heggs, for then they are e/gg/ asperated.

The reaction of young Archie to this donnish joke can only be guessed.

Although natural history was his early love, and one that would persist throughout his long life, Archie became interested in history through the efforts of a master who, Garrod much later recalled, 'so taught us history that the reading of history has been a private hobby ever since'.[6]

In January 1873, at the age of fifteen, Archie entered Marlborough College in Wiltshire as a member of Littlefield House.[7] The head of the house, James Gilmore, was an extraordinarily able mathematician who had obtained a First in the list of Senior Moderators at Cambridge, a distinction corresponding to that of Senior Wrangler at Oxford. Gilmore was also one of the best-loved masters of his day. In a memorial notice he was called 'a reserved man of extraordinary uprightness and kindness', words that could well have been used later to describe Archibald Garrod.[8]

Marlborough College had been founded only thirty years before, but it had already gained recognition as a school that provided a sound education. English public schools at the time placed considerable emphasis on the classics. A demonstrated competency in classics was necessary for anyone who wished to enter Oxford or Cambridge, and, because the classics dominated the curriculum, the teaching of biology and the physical sciences was all too often neglected. Arthur Benson despaired that public schools were turning out 'well-groomed, well mannered, rational, manly boys, all taking the same view of things, all doing the same things'.[9]

Although his brother Alfred urged him to spend more time studying the classics, and although he himself was later to emphasize the importance of a classical education, Archibald did not find these subjects as rewarding as the sciences. His school report in the summer of 1874 was decidedly cautionary:

He is still weak in grammar and parsing; but his Latin prose is very much improved. He has a good sound sense and a considerable ability for English work, and also a fair amount of general knowledge.

Science: He made excellent progress. Takes great interest in the subject.

The headmaster's remarks were encouraging:

Excellent; without abandoning the Classics, I wish he would continue to do his utmost in science, a subject in which he may certainly win distinction.[10]

Archibald's father, influenced by the boy's adverse school reports, considered him so unpromising academically that he resolved to remove him at the age of seventeen from Marlborough before the completion of his studies and send him directly into business.[11]

Fortunately, Archibald was saved from this fate by his brother Alfred, who helped dissuade their father, and by Frederic William Farrar, the headmaster of Marlborough. Farrar, who had appointed George Rodwell to organize science teaching in 1871, disagreed with the public schools' prevailing preoccupation with the classics, and persuaded Archibald's father to leave the boy in school. Rodwell, with Farrar's support, was Archibald's most influential teacher at Marlborough, and later Garrod would speak with great feeling of the crucial role that Rodwell had played in his education:

In our school days, in the seventies of the 19th century, the position of boys who hoped to go on to Oxford or Cambridge, but who could hardly keep pace with their fellows in the Upper School, was not an enviable one. Even those who could have held their own easily in more congenial studies were in danger of losing heart, of regarding themselves as destined to fail in life's race, and parents and masters were apt to share this view. To some such boys the scanty hours devoted to science were like oases in the desert, and to them a few words of encouragement from headmaster or teacher meant much. But for such kindly help they might well have missed their vocation.[12]

Farrar maintained that the ideal curriculum should emphasize science, mathematics, and modern languages, and he was extremely influential in introducing science into the curriculum of English public schools. Later Dean of Canterbury, Farrar was a polymath, unusually well informed in all branches of learning, including the sciences. Charles Darwin, an omnivorous reader, was so impressed with Farrar's *Language and languages*, in which he described the evolutionary interpretation of linguistics, that in 1866 he proposed Farrar for Fellowship in the Royal Society, to which he was elected soon thereafter. When Darwin died, Farrar was selected to deliver the funeral oration in Westminster Abbey, where he gracefully eulogized the great biologist:

… Charles Darwin will take his place, side by side, with Ray & Linnaeus; with Newton and Pascal; with Herschel and Faraday, — among those who have not only served humanity by their genius, but have also brightened its ideal by holy lives.[13]

Farrar's principal field of academic interest was literature, and he is probably best known for his story of school life entitled *Eric, or little by little*. His enthusiasm for the sciences, however, helped him recognize Archibald's talent. Farrar urged Archibald to make every effort to counteract the negative impression he had made on his classics teachers. But Victorian schoolmasters were obdurately unforgiving, as Archibald's last term report card in classics in 1875 shows (Fig. 13): 'he has not done well. His grammar lessons are rarely known and his construing and Latin Prose are full of blunders.' Nevertheless, Farrar was unwavering in his support, '[Archibald is] A boy who may have a very useful career, but it is clear that science is his forte, and I'd like him to follow this bent almost unreservedly.'[14] Which is precisely what Archibald did.

Archibald's interest in natural history became firmly rooted at Marlborough, and extended into the chemical and physical sciences. Two months after entering, he was elected to the school's Natural History Society, a vigorous and energetic organization run by the Reverend T. A. Preston, who had started the Society in 1864. Society records

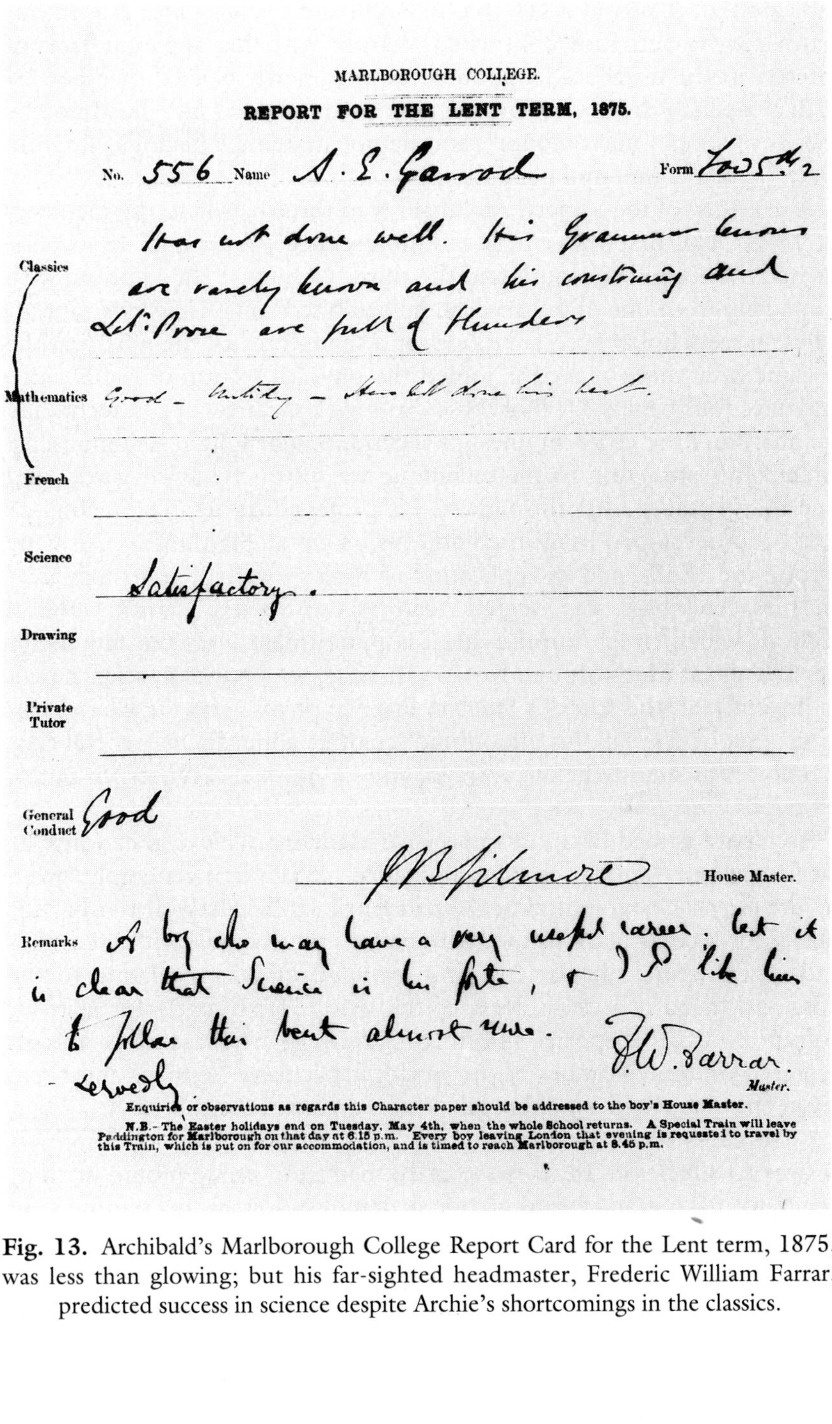

MARLBOROUGH COLLEGE.

REPORT FOR THE LENT TERM, 1875.

No. *556* Name *A. E. Garrod* Form *Lower 5th 2*

Classics *Has not done well. His Grammar lessons are rarely known and his construing and Lat. Prose are full of blunders*

Mathematics *Good — Untidy — Has all done his best.*

French

Science *Satisfactory.*

Drawing

Private Tutor

General Conduct *Good*

_____ *J. S. Gilmore* _____ House Master.

Remarks *A boy who may have a very useful career, but it is clear that Science is his forte, & I should like him to follow this bent almost unre-servedly.*

F. W. Farrar.
Master.

Enquiries or observations as regards this Character paper should be addressed to the boy's House Master.

N.B.— The Easter holidays end on Tuesday, May 4th, when the whole School returns. A Special Train will leave Paddington for Marlborough on that day at 6.15 p.m. Every boy leaving London that evening is requested to travel by this Train, which is put on for our accommodation, and is timed to reach Marlborough at 8.45 p.m.

Fig. 13. Archibald's Marlborough College Report Card for the Lent term, 1875, was less than glowing; but his far-sighted headmaster, Frederic William Farrar, predicted success in science despite Archie's shortcomings in the classics.

indicate that it played a key role in Archibald's school life. His enthusiasm for its activities contrasted sharply with his apparent lack of interest in the usual academic subjects. As a newly elected member, he read a paper to the Society on ' "The marvels of pond life, illustrated by means of several microscopes", and demonstrated a Chladni's vibrating plate[15] and a compound pendulum'.

A mainstay of the Society, Archibald read three papers to the members in 1874. The first was entitled 'Fluorescence', illustrated by experiments, the second 'Colour', and the third 'Comets'.[16] In addition, with some help from one of his masters, he exhibited some Geissler's tubes.[17] These were scholarly presentations for a schoolboy. In the next year, he became even more active; he joined the physical section of the Society, and gave four papers. Two of these dealt with spectroscopy (a technique for analyzing the different lines, or spectra, of which light is composed). Archibald's attraction to the technique was distinctly avant-garde, and one that would persist throughout his professional life. The technique had been developed by Bunsen and his group at Marburg in Germany during the 1860s, and its application to biology was still in its infancy.[18] In July, Archibald was elected head of the Society's entomological section, which to his surprise and disappointment attracted few members. While at Marlborough he won three general prizes in science and, in his last year, the school's Stanton Prize in physics and the Clark Prize in geography. Despite his undoubted scientific aptitude, his general class standing was slightly below average, and he finished seventeenth out of 29 pupils.[19]

Archibald gained his most impressive academic achievement when he sat for the inter-public-school examinations in physical geography given by the Royal Geographical Society in April 1875. He won the bronze medal for second place in the examination, which dealt with volcanoes and other natural phenomena, and included questions relating to the flora and fauna of various parts of the world. Fortunately for Garrod, volcanoes were a special hobby of Rodwell, whose travels abroad centered on the volcanoes of the world, particularly Vesuvius and Etna; given any chance, he would regale his students on the subject.

Candidates had to answer 12 of 18 questions. Although Archie's answer to Question 14 shows that he had not yet overcome his tendency to make grammatical errors, it demonstrates his deep interest in natural history. The question read:

The assemblages of living creatures found in two distant parts of the earth generally differ from one another, and those found in two neighbouring places generally resemble one another. What is supposed to be the cause of such differences and such resemblances? Are there exceptions to this rule, and, if so, how do they occur?

Archibald's answer and spelling are given verbatim:

The cause of the differences between the animals of different parts of the world is that the speceimins from which they both originally sprang have owing to the difference of climate and physical features of the countries which they inhabit been subject to different influences and those which were *"fittest"* to survive in one country are not so in another, and by the survival of the fittest diverent forms have been preserved in the two regions, and the species have become more distinct from generation to generation, and eventually they appear to bear no resemblance to each other, whereas in neighbouring tracts of country the climates & c generally resemble each other and consequently no very marked difference grows up between the two branches.

For instance a species might be able to live in two countries adjacent as India and Burmah which are not calculated to live in South America and so on.[20]

Archibald's examiner, commenting on the papers, wrote, 'The bronze medallist was very near the gold medallist in the general paper, which was the most difficult.' Archibald apparently did less well on the second paper in the examination, which was devoted to China.

Although Archibald's academic accomplishments during his years at Marlborough were far from spectacular, they pointed him firmly toward a career in science. There can be little doubt that the following tribute, excerpted from Garrod's address to the Osler Club more than fifty years later, refers to his years at Marlborough:

Each one of us, when he casts his thoughts backwards over his own life, recalls a few people who have influenced him powerfully either for good or evil; but it is upon the former that our memory likes to dwell. We recall the schoolmaster who awakened our interest in what has been our life's work; he who by a word of encouragement spoken in good season helped to give us confidence, and showed us that one person at least believed that there was something in us. Another may have succeeded in interesting us in subjects for which we had no aptitude — classics perhaps — and who seems to walk beside us when we tread the streets of Athens or of Rome. No doubt as we grow older these people tend to grow in stature in our memory, but they undoubtedly possessed striking personality, and sowed a grain of mustard seed which has produced a tree.[21]

A Career on the Move: Oxford and London

On 19 January 1877, aided by his improved academic performance in his last year at Marlborough, and bolstered further by his having spent the academic year 1875–6 attending a special course in chemistry at University College, London, Archibald Garrod entered Christ Church at the University of Oxford as a Commoner.[1] His father did not want him to go to Cambridge, for he was convinced that Archibald would be compared unfavorably to his brother Herbert, who had been a stunning intellectual success. Archibald had a brief setback at the beginning of his university career when he failed to pass the compulsory examination in Holy Scripture; but he overcame this necessary but vestigial educational hurdle in his second term, and went on to obtain a first-class degree in the School of Natural Science (Chemistry) in June 1880.

While at Oxford, Archibald kept in touch with his family, and although his letters home were rather infrequent, he always wrote enthusiastically. On 27 April 1880 he wrote to his mother:

Many thanks for your letter. I am afraid I have not written home very often lately. I hope you will excuse it partly because I am so busy just now, and partly because being busy I have very little to say.

Although Archibald was not much of an oarsman and never rowed for his college, he had enjoyed rowing since his prep-school days, and this carried over to his years at Oxford. In the same letter to his mother he wrote:

The eights begin on Wednesday May 5. Christ Church was expected to have a very good boat this year, but although we have the Varsity Stroke I am afraid it has not come up to the expectations formed of it.[2]

On 25 October he wrote home of his concern that he had not heard much from his favorite sister, thirteen-year-old Effie, particularly as he had recently sent her a small present which she had not acknowledged. He also wrote that he had been invited to lunch with Ray Lankester, the distinguished biologist who would later support him for fellowship in the Royal Society, and had attended a lecture on 'speech' by Alexander Graham Bell.[3] Since Bell had introduced the telephone to the world only four years earlier, his lecture at Oxford was an event not to be missed.

Conscious that his father still had reservations about his intellectual ability, Archibald worked hard at his studies. He also clearly enjoyed himself, particularly his tutorials and his laboratory work in chemistry.

Archibald had the good fortune to have A. G. Vernon Harcourt, Lee's Reader in Chemistry, as his tutor. Harcourt was a brilliant chemist who had been elected to the Royal Society at the age of twenty-nine. Although he made significant contributions to scientific research, particularly in the field of physical chemistry, his students knew him as a devoted teacher who 'lived in the lives of his students'.[4] Harcourt insisted that students write with clarity, and this influence showed plainly in all that Garrod wrote in later life. He quite overcame his schoolboy difficulties with spelling and punctuation, and became a precise and graceful writer.

Few details exist to suggest any lasting friendships that Archibald made while at Oxford. The Dean of Christ Church at the time was Henry George Liddell, joint compiler with Robert Scott of the classic *A Greek–English lexicon*. C. L. Dodgson, who immortalized Liddell's daughter Alice under his pen-name Lewis Carroll, was tutor of mathematics. There is no evidence that Garrod was taught by any of these luminaries; nevertheless, nourished by the rich intellectual environment of Oxford, young Archibald decided to embark on a career in science.

In 1879 Archibald won the Johnson Memorial Prize, offered every four years for an essay on 'some Astronomical or Meteorological subject'. The subject that year was 'The History of the Successive Stages of our Knowledge of Nebulae, Nebulous Stars and Star-clusters, from the time of Sir William Herschel'. The detached orderliness, neatness, and rigor of Archibald's essay presaged the systematic and original approach he would bring to solving medical and scientific problems:

In these days of rapid scientific progress there is a tendency to accept the facts of nature, as at present known, without glancing back at the slow and difficult stages by which the knowledge of these facts has been arrived at. Yet such a retrospect is by no means unprofitable, since it warns us that hasty generalizations upon insufficient data retard rather than advance the progress of knowledge, and that the theories of the day must not be accepted as necessarily expressing absolute truths.[5]

Spectroscopy was an important scientific tool for investigating the heavens, and Garrod's earlier excursions into spectroscopy at school undoubtedly helped in the success of his essay. A few years later, his knowledge and mastery of this important scientific tool became pivotal in his chemical investigation of colored urine.

Archibald's years at Oxford, capped by his first-class degree in natural science, gave him a sound basis for a scientific career. His academic success surprised and reassured his father. Harcourt had encouraged Archibald to pursue a career in chemistry, but, perhaps because the elder Garrod had already shown that a medical career could be greatly illuminated by the application of science, Archibald decided to follow his father into medicine.

He returned to London in 1880 to live with his parents at 10 Harley Street, and entered the Royal Hospital of St Bartholomew, which was one of the oldest and most distinguished hospitals in London.[6] Founded in 1123, it boasted among its distinguished physicians William Harvey, whose discovery of the circulation of the blood is often used to date the birth of modern medicine. Like his father, Archibald had a strong historical sense of the importance of the old teaching hospitals in the provision of medical care to the poor in London, which may have influenced his selection of St Bartholomew's Hospital, affectionately known as Bart's, for his clinical studies. Moreover, his natural shyness might have deterred him from entering King's College Hospital, where his father played such an influential role.

When Archibald entered Bart's, it was to learn the art and practice of medicine. However, a laboratory for practical chemistry had been built in 1866, because the subject had become too complex to be taught by clinical assistants. Until the middle of the nineteenth century, medical school offerings were limited to surgery, dissection, and anatomy. It was not until 1890, ten years after Archibald entered medical school, that Bart's provided a bacteriological laboratory. By the turn of the century, laboratories for the study of biology, physiology, and physics were added; but it would be a long time before the teaching of science in the London medical schools would approach that available at Oxford and Cambridge.[7]

Garrod recalled these early London teaching hospitals in a 1930 address to the Abernethian Society, the student research society founded in 1795 and later named in honor of surgeon-educator John Abernethy.[8] Garrod reminded his young audience that when he entered medical school fifty years earlier, there was no electric light and no telephones. Most of the patients who came to the hospital lived in homes with no fixed bathtubs, and public baths, dotted throughout the city, were few. The automobile had not been invented; when a physician was called at night to see a desperately ill patient in the hospital, a 'porter was sent in a hansom cab to fetch him'.[9,10]

The large wards, with no private accommodations, were warmed rather ineffectively by open coal stoves. Infectious diseases were common, and the wards were filled with patients with typhoid, diphtheria, scarlet fever, and occasionally typhus. The doctors wore the mandatory frock-coat and tails that bespoke the successful Victorian physician. As they walked the wards, accompanied by a train of medical students, their only diagnostic aids were a thermometer, recently introduced by Clifford Allbutt[11], and a wooden monaural stethoscope.

Although the 'new' medical school had opened the year before Archibald entered Bart's, the adjunctive teaching space remained the Museum and the Library. The two laboratories, one for pathology and one for physiology, were still empty. Bacteriology did not yet exist as an aid to diagnosis, although Louis Pasteur had pointed the way in 1863. Robert Koch would not identify the bacillus that caused tuberculosis until 1882, three years after Archibald's brother Alfred Henry had died from that disease. The heroes of medical practice were those physicians interested in metabolism: Thomas Addison, Samuel Wilks, and William Gull, all of London's Guy's Hospital. Autopsies were common, but histology was in its infancy. Edward E. Klein, a brilliant German-trained histologist who later would propose Garrod for the Royal Society, had just joined Bart's staff. Botany formed an important part of the curriculum, and elegance in the dubious exercise of polypharmacy was highly prized.

In 1880 the consulting staff at Bart's was a roster of capable and distinguished medical men; but it was a sign of the times that many of the eminent physicians Archibald encountered during his student years were extremely conservative and antiscientific. One of them, William Church (later Sir William), a physician of considerable inherited wealth, accepted the presidency of the Royal College of Physicians, despite his frequently expressed preference for the life of a country squire to that of an over-worked medical consultant.

Another physician on the staff Samuel Jones Gee, differed greatly from Church. Steeped in the classics and desperately shy, Gee had few social graces and avoided the limelight. Despite these characteristics, he was a superb, if formidable, clinical teacher, who hated pretentiousness and who would acknowledge only those physical signs that were beyond dispute. He was noted for the pithiness of his remarks, and once said on the subject of rectal feeding: 'A very little food in the stomach is better than a large amount in the rectum.'[12] Garrod admired him greatly, and was later to call him 'a modern Hippocrates'.[13] Gee was the co-discoverer in 1888 of the clinical condition which he called 'the Celiac affection', but which is now usually called non-tropical sprue.[14] Withal, Gee was militantly antiscientific. He went so far as to refuse to deliver the Harveian Oration of the Royal College of Physicians because he heartily disapproved of William Harvey's injunction to 'seek and study out the secrets of Nature by way of experiment'.[15] He modified this extreme position somewhat as he grew older, and in his Lumleian Lecture in 1899 he admitted to the usefulness of a scientific approach to medicine.[16]

Other distinguished physicians on Bart's staff included Reginald Southey, an able man without ambition, and James Andrew. Southey was the inventor of 'Southey's tubes', hollow tubes inserted into the legs for the relief of dependent dropsy, a useful, if barbarous technique, that continued to be employed sporadically for the next seventy-five years. According to Garrod, Andrew introduced him to the importance of cultivating the art of clinical medicine. A kind-hearted and shy man, Andrew wrote little and was not among London's medical élite; but Garrod praised him as a wise physician, who taught 'by precept and example'.[17]

It was fortunate for Archibald, fresh from Oxford and eager to follow the science of medicine, that although many of the physicians on the staff had little time for science, there were a few notable exceptions. The Department of Chemistry was led by William J. Russell, who, in addition to being a first-class chemist and teacher, was interested in public health. This latter activity had been fostered by Russell's teacher, Edward Frankland, who, while contributing to an understanding of the laws of chemical valency, did fundamental work on water pollution.

Pollution was a serious concern, and the London of Garrod's time was one of the most disagreeable European cities in which to winter. The ever-present fog exacerbated the respiratory ills of anyone breathing the foul and polluted atmosphere of the city. In Dickens's telling phrase, smoke was 'the London ivy' that wreathed itself around every dwelling and clung to every building. Russell, in an effort to establish a causal connection between fog and disease, studied its composition and physical properties. On one occasion when dense fog enveloped the city, Russell found that samples of air taken from several sites around central London had more than four times the normal concentration of carbon dioxide.[18]

Archibald's intellectual abilities blossomed at Bart's, and he had a distinguished student career there. Soon after his arrival, he took time in the evenings to reconstruct and rewrite his Johnson Memorial Prize essay on astronomy, and arranged for its private printing in Oxford in 1882. He inscribed a copy for his father and another for Laura Elisabeth Smith, his wife-to-be. It seems doubtful that the eighteen-year-old Laura was as excited by the 44-page astronomical treatise as was Archibald's father; but, although he inscribed no felicitous message, there is little doubt that the shy Archibald Garrod was moving tentatively in the direction of marriage.

In 1881 Archibald won the Junior Scholarship Prize, and in May 1884, the Brackenbury Scholarship in Medicine, a scholarship that carried with it a small sum of money. He celebrated his prize by visiting Norway in the summer with a fellow student Henry Lewis Jones. Jones was a Cambridge graduate three years ahead of Archibald. He too had obtained a First in the natural sciences, and he too was later also to apply his scientific knowledge to medical practice. Jones was an able physicist, as well as a good medical student, and he was to become head of the Electrical Department of Bart's. By his caution and reliance on basic science, he gave this potentially dubious discipline a scientifically rational basis. Jones paid particular attention to the therapeutic effects of various forms of electrical therapy on patients with rheumatic and other ailments, partly, one suspects, because of his friendship with Archibald Garrod. Jones's book, *Medical electricity*, written in collaboration with W. E. Steavenson and first published in 1892, was a pioneering and regularly updated volume in the field.[19] The seventh and last edition, published in 1918, included a full discussion of the growing use of X-rays as an adjunct to the surgical treatment of cancer.

While on holiday in Norway, Garrod and Jones visited the leper colony at Bergen, spending time with the patients and talking to the physicians who looked after them. On his return home, Garrod gave a paper on his experiences to the Abernethian Society. He undoubtedly startled his audience when he casually said that there were 6000 lepers in Norway, and that 'the Western coast of Norway now vies with Crete for the unenviable reputation of being the chief remaining stronghold of the disease' — an arresting statement that later yielded to more extensive research.[20]

Garrod's paper, written while he was still a student, provides an early hint that he would become more than a mere chronicler of disease. In addition to the clinical description of leprosy, Garrod described the geography of western Norway and discussed the history of the disease throughout the world. A theory commonly held at the time was that leprosy was caused by eating fish. Although Garrod seemed to favor the notion, he was bound to admit that he found the evidence in support of it far from convincing. The fish theory had powerful and influential critics. Sir William Moore was convinced that the theory could not apply to leprosy, at least as seen in India. He pointed out that leprosy was unknown among the Kashmiris, who were well-known to be avid consumers of fish — fresh, dried, and salted. In contrast, leprosy was commonly seen among the Goojuns, another Indian group who abominated fish.[21] Four years after Archibald visited Bergen, Moore's view was beginning to prevail. Hercules MacDonnell wrote in the *Lancet*, 'Nowhere [in Norway] did I perceive that any credence was attached to the fish origin theory.'[22] Garrod had been characteristically prudent in his report, and while acknowledging the fish theory and perhaps tending toward it himself, he awaited additional evidence before accepting it as the cause of leprosy: 'Although it [the fish theory] is not able to explain all the facts, it is the most satisfactory of those hitherto advanced.'[23]

Although Garrod was not ordinarily gregarious by nature, he thoroughly enjoyed his years as a student at Bart's and the friends he made there. One with whom, in particular, he competed throughout his student days for the several medical prizes was Samuel Habershon. Habershon edged out Garrod for the Gold Medal in medicine, surgery, and midwifery, as well as the Kirkes Scholarship and Gold Medal for clinical medicine. He later became friend and physician to the Prime Minister William Gladstone. Garrod and Habershon were considered so inseparable by their fellow students that they were often referred to as 'Harrod and Gabershon'.

In 1884 Garrod passed his qualifying examination in medicine for membership of the Royal College of Surgeons (MRCS). He was now licensed to practise medicine. The following year, he obtained from the University of Oxford the additional degrees of Bachelor of Medicine and of Surgery (B.M., B.Ch.). Although the additional Oxford degree was not strictly necessary for the practice of medicine, for those, like Garrod, who were intent on an academic career it was a near-requisite.

4

The Postgraduate Years

IN Garrod's time, a *Wanderjahr* for those who had achieved distinction in medical school was extremely common. Encouraged by his father, Garrod spent the winter months following his graduation in Europe. He had the best of reasons; science on the Continent, particularly medical science, was in the middle of an extraordinary decade of discovery. In France, Louis Pasteur had developed a vaccine against anthrax in 1881. The next year, in the first definite association of a germ with a disease, Robert Koch discovered the bacterium that caused tuberculosis. A year later, Koch discovered the cholera bacterium, *Vibrio cholerae*, and proved that this devastating disease could be spread by food and water. In the years after Garrod's graduation, Ilya Illich Metchnikoff had discovered that phagocytes, the white cells of the blood, could attack and destroy invading organisms. Pasteur was on the threshold of developing a vaccine against rabies, and had saved the life of a young boy Joseph Meisler, bitten by a rabid dog.

The Allgemeines Krankenhaus in Vienna was regarded as one of the best in the world. Indeed, so many American students came that they founded a new organization, the American Medical Association of Vienna. If Garrod was to become a leader in the medical profession, it was almost required that he go to Vienna and learn from the masters. He decided to further his education at the famed medical clinic in Vienna, as had Sir Wilmot Herringham, a senior colleague at St Bartholomew's, before him. Garrod learned German and established friendships and medical contacts that would prove extremely useful in his future academic career. In the German clinic system, a professor, accompanied by experienced assistants, instructed medical students in a series of lecture-demonstrations. Garrod attended several medical clinics, as well as the lecture-demonstrations of two of the leading European laryngologists of the day, Johann Schnitzler and Leopold Schrötter von Kristelli. He returned home full of enthusiasm for the clinic system, and some thirty years later recounted its advantages in his testimony before the Haldane Commission on Medical Education.

Garrod was always interested in diagnostic problems, and regarded laryngoscopy as a tool for improving the accuracy of clinical diagnosis. Physicians trained in the technique and interpretation of laryngoscopy could determine if hoarseness were due to a polyp of the larynx or to tuberculosis. Schnitzler and Schrötter had made Vienna the leading center for the study of laryngology, and students from all parts of the world flocked to their clinic. Garrod's friend from Bart's Samuel Habershon, accompanied him to Vienna, and they both registered for two courses in January and February 1885. Although trained in clinical medicine, Schnitzler and Schrötter believed laryngology could only develop fully if it were given major status in the medical school, and it was due to Schrötter that the first university clinic of laryngology was established in Vienna. Later, when Garrod argued that pediatrics should be a recognized medical specialty, he was no doubt influenced by his experience in Vienna.

On his return to England, in the spring of 1885, Garrod capitalized on his experience in Vienna and on his scientific and literary leanings by writing *An introduction to the use of the laryngoscope*.[1] Laryngoscopy was far less advanced in England than on the Continent, and Garrod's book was soon out of print. Shortly after his book was published, Garrod encountered a patient with paralysis of the vocal cords, and was able to use his recently acquired laryngoscopic skill in following the progression of the disease. Although Garrod regarded laryngoscopy simply as a tool that permitted a more precise diagnosis, he was held in sufficiently high regard by the leading London laryngologists to be elected vice-president of the laryngoscopic section of the Royal Medical and Chirurgical Society.

Early in 1885 Garrod was appointed House Physician to Dyce Duckworth (later Sir Dyce). Duckworth, a long-time friend of Garrod's father, specialized in gout and the rheumatic diseases. He had taken a great liking to young Garrod while Garrod was still a medical student. Garrod was delighted at the appointment, even though Duckworth was a conservative clinician of the old school. It was said that Duckworth 'opposed in no uncertain terms such developments as the emancipation of women, the relaxed observance of Sundays, and modern trends in art, music, dancing, and fashion'.[2] He had been appointed Physician to St Bartholomew's in 1883, and he emphasized at every opportunity the canonical imperative that physicians should above all else cultivate the art of medicine. Although Garrod was fully aware of Duckworth's hostility to change in the practice of medicine, he also knew that if he worked with Duckworth he would be assured not only of a rigorous clinical training, but, equally important, of powerful academic support in the years ahead provided he did well.

As part of Garrod's clinical training, Duckworth insisted that he see as many patients with as many different diseases as possible. A prolific and thoughtful contributor to medical literature himself, Duckworth encouraged Garrod to publish case reports on patients who demonstrated some special feature of a disease. During his twelve-month apprenticeship, Garrod published several papers, including a neurological paper describing four cases of sclerosis of the spinal cord.

Garrod was set on becoming a consultant physician, and in order to accomplish this goal he had first to pass advanced examinations in medicine. These academic hurdles proved no obstacle. His able and orderly mind enabled him to pass the membership examination of the Royal College of Physicians in 1885, and to receive his doctorate of medicine from Oxford the following year, with a thesis on rheumatoid arthritis.

Early in 1886, with his career on a firmer footing, Archibald Garrod became engaged to Laura Elisabeth Smith (Fig. 14). Her father, Sir Thomas Smith, Bart., a friend of Archibald's father, was one of the most beloved and successful surgeons in London. Smith was in the habit of inviting the up-and-coming house officers at Bart's to tea at his home on Stafford Place. It was at one of these tea parties that Archibald first met Laura. Some years later, while punting on the tranquil waters of the Thames with a rose romantically affixed to the bow of the boat, Archibald proposed marriage and Laura accepted. After a short engagement, they were married on 27 May 1886, in the Parish Church of St Thomas, Portman Square, London. Archibald had just turned twenty-nine, and Laura twenty-two. Archibald's parents and Laura's father were witnesses at the wedding ceremony (Laura's mother had died seven years earlier). It seems certain that the ceremony was attended by many medical and surgical luminaries of London, as well as by Laura's and Archibald's friends, and it is also likely that Archibald's long-time friend Samuel Habershon, by now a colleague at Bart's, was best man at the wedding.

When Duckworth learned of Garrod's engagement, he wrote warmly to his increasingly promising protégé:

March 31 1886

My dear Garrod,

Just as I sit down to write this note, I have yours.

I am glad to know that you have had a happy and useful term of office. Certainly nothing could be pleasanter than our personal relationship during the past twelve months. I have always felt that I had a very conscientious, sympathetic, and skilful lieutenant in charge of my patients.

Fig. 14. Laura Elisabeth Smith, Lady Garrod.

Now you will have time to think of something else than physic for a time, and you will find such *lacunae* come none too often in after life. Hence, I would say — "Profitez vous de l'occasion". You want a change and some repose. With this note, I send you a small wedding present from Mrs. Duckworth and myself which will serve to remind you and your "future" of our high regard for, and interest in, you both. With all good wishes, including one that you may someday become again connected with me as a colleague in Smithfield [the area of London in which St Bartholomew's Hospital is located], believe me to remain

Yours very sincerely,
Dyce Duckworth.[3]

Although repose was never high on Archibald's list of priorities, he took Laura to Norway on their honeymoon, returning to territory remembered from his trip with Henry Lewis Jones two years earlier. On their way to Norway, they stopped at the Yorkshire seaside town of Scarborough, where Garrod had spent many happy summer holidays as a child. The holiday season had not yet begun, and Archibald, never the profligate, was pleased to write to his parents that he had found 'inexpensive, clean lodgings' for himself and his bride.[4]

The young couple spent nearly two months traveling throughout Europe. Archibald planned a honeymoon that would relive the joys of his first visit to Norway and introduce Laura to the fjords and seascapes of that country. They went first to Molde, where Archibald had spent part of his summer holiday with Jones. They then traveled by train to Vossevangen and by carriage over the scenic and dramatic dirt road from Stalheim to Gudvangen. They continued their journey by boat to Naerøyfjord and Aurlandsfjord, and then on to the much larger Sognefjord and Vadheim. Garrod and Laura probably traveled by horse from Innvik to Faleide. They journeyed also to Bergen, where Archibald had visited the leper colony as a medical student, and then went to Stockholm, which he found, 'As charming as ever and the steam-boating delightful.'[5]

Relaxed and happy, Archibald and Laura walked, took photographs, attended church, and were enraptured by the glorious scenery of the fjords. Archibald wrote to his mother, 'Laura is wonderfully well ... we both feel very much set up by the change.'[6] Laura shouldered most of the responsibility for family correspondence, but Archibald continued to write to his parents and his brother Herbert. His mother wrote reassuring him that she had paid his subscription to the weekly *Times* so that he could keep abreast of English news — even on his honeymoon, Archibald did not want to lose touch with London life. He had learned that Oliver Wendell Holmes, approaching his eightieth birthday, had visited the Garrods; and he wrote from Stockholm expressing regret that his sister Effie had been away and missed seeing the great man. While on their honeymoon, Archibald and Laura received the good news that Archibald's father had received his knighthood from Queen Victoria. The knighthood probably came as no surprise, for Archibald must have known that his father had been approached earlier in the year to enquire whether he would accept such an honor.

Homeward bound, Archibald and Laura stopped briefly at Copenhagen before journeying through Germany to Lübeck, Hamburg, and Cologne. Finally, reaching the port at Flushing, they caught a boat to England, and arrived home on 18 July. Their honeymoon had provided a romantic beginning to their marriage; but one has the impression that Archibald, particularly toward the end of the trip, was impatient to return home and get on with his life's work.

Duckworth knew what he was talking about when he warned young Garrod that moments of repose were few and far between in the life of a physician. From the time he returned home from his honeymoon and established residence at 9 Chandos Street in Cavendish Square, Garrod's professional life would prove increasingly busy and demanding. Laura, by nature more outgoing than Archibald, was content to provide a warm and comfortable home for her husband and herself. Although she had received little formal education, she had a lively mind and read widely. She enjoyed *The Times* crossword puzzle, was an ardent bridge-player, and embroidered curtains and intricate tapestries, particularly for large pieces of furniture. On 28 December 1887 their first child was born: Alfred Noël, named in honor of Archibald's father, and born sufficiently close to Christmas to be called Noël.

Garrod was eager to continue his medical training at Bart's, and in 1887 he was appointed Casualty Physician, a post he held for two years. Robert Bridges, who later became England's Poet Laureate, and who had preceded Garrod as Casualty Physician, has given a graphic description of the atmosphere in the clinic in Garrod's time:

No description could do justice to the strange hubbub in which the auscultation had to be carried out. The rattle of carts in the street, the hum of voices inside, the slamming of doors, the noise of people walking about, the coughings of all kinds, the crying of babies, the scraping of impatient feet, the stamping of cold ones, the clinking of bottles and zinc tickets; and after eleven o'clock, the hammering of and tinkering of the carpenter and blacksmiths who came not unfrequently at that hour to set things generally to rights.[7]

It was in the environment of the clinic, so vividly described by Bridges, that Garrod would get his first taste of what it meant to be a doctor (Fig. 15). It was not, however, a good place to learn clinical medicine. The atmosphere was too hectic, and there were no means of diagnosis except clinical examination. The physician's principal role was to determine whether the patient was well enough to go home or sufficiently sick to be admitted to a medical or surgical ward. Garrod's position was not full-time, and for several years he saw private patients in Chandos Street. Archibald also helped his father in his busy consultant practice around the corner in Harley Street, where the elder Garrod specialized in the rheumatic diseases. In 1886, Archibald was elected to membership of the Royal Medical and Chirurgical Society, as his father had been 32 years earlier. Young Garrod greatly prized the collegial atmosphere that the Society provided. Often accompanied by his father, he regularly attended the monthly meetings, served on committees, including the publication committee, and frequently presented papers

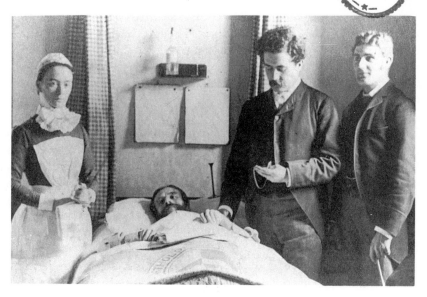

Fig. 15. Garrod on ward rounds at Bart's, *circa* 1903.

on rheumatological and other clinical subjects that were subsequently published in the Society's *Transactions*.

Although Garrod's scientific curiosity never left him, he heeded Duckworth's advice, and became first of all a well-rounded clinician. A rewarding opportunity to extend his clinical experience arose in 1888, when he was appointed Assistant Physician to the West London Hospital, where he remained for eight years. In 1888, he was elected a Fellow of the Royal College of Physicians only three years after obtaining membership of the College — an unusually short interval. In 1889 he was appointed Physician to the Marylebone Clinic — a less prestigious post, but one that presumably offered additional opportunities for clinical work.

In 1894, in association with two senior colleagues, Wilmot P. Herringham and William J. Gow, Garrod wrote a small handbook of medical pathology to aid students at Bart's as they studied morbid anatomy in the hospital museum.[8] In December 1895, he received two additional appointments, which he held concurrently for six years. These were a demonstratorship of morbid anatomy in the medical school and a medical registrarship in the hospital, where he taught students clinical medicine at the bedside as well as in the out-patient department. Between 1896 and 1901, he and his colleagues in the out-patient department took part in the annual preparation of *Statistical tables of the patients under treatment at St Bartholomew's Hospital*. Though the task must have been somewhat tedious, Garrod undertook it with his usual conscientiousness.

Unlike most of his colleagues, even those considerably his senior, Garrod was already well known in Germany and France, and many friends and medical visitors came to Bart's to see him. In 1896, he translated *A treatise on cholethiasis*, written in German by his friend Professor B. Naunyn, for the New Sydenham Society (named in honor of the distinguished eighteenth-century English physician Thomas Sydenham).[9] In 1897, Garrod visited Germany to deliver a paper on hematoporphyrinuria.[10] He also translated for the New Sydenham Society two books, one in French and one in German, on yaws, a non-venereal disease caused by the spirochete *Treponema pertenue*, an organism that was commonly encountered in tropical Africa, the West Indies, and the Far East.[11,12]

Although it will be convenient in tracking Garrod's subsequent professional career to divide his interests into those that were primarily clinical in nature and those that were more strictly scientific, one should recognize that his clinical and scientific work were carried out at the same time. The eight years he spent at the West London Hospital enabled him to expand his general medical knowledge and to gain clinical experience working in the out-patient department. In 1894 he became, probably because of his long interest in rheumatoid arthritis, Physician to the Alexander Hospital for Diseases of the Hip, a position he held for only two years. He resigned from both hospitals in 1896, at a time when he became increasingly absorbed with the spectroscopic analysis of urine. During these years Garrod published a variety of articles in medical journals, particularly on rheumatic diseases and gout, as well as pursuing simultaneously his scientific interests.

Garrod's papers, both clinical and scientific, were written with a clarity and grace that drew admiration from his colleagues. He took immense pleasure in writing, and always sought the felicitous word and appropriate phrase, in sharp contrast to the stumbling writing of his childhood. His small, at times minuscule, handwriting was always legible — a far cry from the stereotype of a physician's hand.

When Garrod published his first major work, in 1890, *A treatise on rheumatism and rheumatoid arthritis*,[13] he gave an inscribed copy to his father: 'To my dear father from his affectionate son, Archibald E. Garrod' (Fig. 16). The book received wide acclaim, and became an instant success. It was translated into French the following year.

To my dear father
from his affectionate Son.
Archibald E Garrod

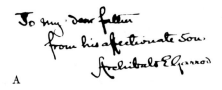

A

TREATISE ON RHEUMATISM

AND

RHEUMATOID ARTHRITIS.

BY

ARCHIBALD E. GARROD, M.A., M.D., Oxon., M.R.C.P.,

ASSISTANT-PHYSICIAN TO THE WEST LONDON HOSPITAL;
LATE CASUALTY PHYSICIAN TO ST. BARTHOLOMEW'S HOSPITAL.

„Das Wenige verschwindet leicht dem Blick,
Der vorwärts sieht, wie viel noch übrig bleibt."
GOETHE (*Iphigenie auf Tauris*, i. 2).

With Charts and Illustrations.

LONDON:
CHARLES GRIFFIN AND COMPANY,
EXETER STREET, STRAND.
1890.

[*All rights reserved.*]

Fig. 16. This copy of *A treatise on rheumatism and rheumatoid arthritis*, inscribed 'To my dear father from his affectionate son', must have greatly pleased Sir Alfred, whose own career involved extensive study of rheumatology.

Garrod's interest in rheumatic diseases was not exhausted with the publication of this *opus magnum*, and for more than a decade he wrote intermittently on rheumatological topics. In these papers, Garrod did not mention any possible connection between infectious agents and rheumatoid arthritis; instead, he supposed the disease to have a neurological basis — a common view at the time.

One such paper, 'An attempt to determine the frequency of rheumatic family histories amongst non-rheumatic patients', co-authored with E. Hunt Cooke and published in the *Lancet* in 1888, revealed Garrod's developing interest in the influence of heredity on disease.[14] While Garrod was clearly convinced that hereditary influences play an etiological role, he was very concerned that the way in which the patients had been collected for the study might have introduced a bias. In his closing paragraph, Garrod wrote:

To sum up, these statistics indicate that, whereas about 20 per cent. of the patients who present themselves at London hospitals suffering from morbid conditions which stand in no recognized relation to rheumatism have family histories of rheumatic fever, amongst those who have themselves suffered from rheumatism or allied diseases, such histories are obtained in some 30 to 35 per cent.; in each instance, however, considerable allowance must be made for erroneous information.[15]

In November 1892 Garrod was appointed, to his great delight, an Assistant Physician to the out-patients of the Hospital for Sick Children in Great Ormond Street. This was the first hospital in the United Kingdom to be devoted entirely to the care of children, and it had opened in 1852 with 10 beds. Charles Dickens had given the hospital national publicity by writing a number of articles on the appalling plight of sick children in London, emphasizing the serious inadequacy of existing care, and urged the public to make contributions to ensure improvements. In honor of Queen Victoria's Jubilee the sum of £5910. 3*s*. 11*d*. had been collected throughout the British Empire, enabling the hospital to expand the number of beds to 240.[16]

Although Garrod was not much interested in the infectious diseases, diarrheas, and inanition that were most common among patients in the hospital, his colleagues admired him greatly for his academic approach, wisdom, and honesty of purpose. But even the greatest admirers of this handsome man of medium height and build, with a fine head and piercing yet kindly eyes, could not but admit that his entry on the ward, with his deliberate gait, and his slightly reserved but ever-courteous manner, seemed largely ceremonial. His ward visits would usually center around the test-tube stand, where his dissertations upon rare and obscure medical problems in childhood were far beyond the capacity of his audience. It seems clear that the new concepts in medicine and the scientific way of looking at human disease represented by Garrod were lost on his audiences. He seemed to them obsessed with rarities — bizarre inherited metabolic disorders of no real interest to the practitioner of medicine.

Garrod soon became widely respected and liked by his colleagues at Great Ormond Street, and in June 1899 he was promoted to Physician to the Hospital, at the age of forty-two. The appointment to Physician was crucial to the development of his scientific career, because it marked the first time that hospital beds were allocated to him. He could admit patients directly under his own care, and no longer had to depend on out-patient visits or on the goodwill of colleagues, who would occasionally refer patients to him. He was immensely happy during his years on the staff of Great Ormond Street — a happiness that was shared by Laura, who regularly donated from her own purse to the hospital, as well as actively engaging in volunteer work among the young patients.

Garrod had little appetite for small talk, and he did not much enjoy the social rounds that often attended physicians engaged in a fashionable consulting practice. Shy by nature, he much preferred to look after children, who took an instant liking to this large, uncomplicated teddy bear of a man with his droopy walrus mustache (Fig. 17). In 1896 Robert Hutchison, who subsequently became one of London's most distinguished pediatricians, was a House Physician at Great Ormond Street. Garrod was still an Assistant Physician at the time, but when the Senior Physicians were away he supervised the care of their patients. On such occasions Robert Hutchison had Garrod as his chief. Hutchison wrote of Garrod:

He was a man whom everyone admired, "a scholar and a gentleman", but he was not one about whom anecdotes catered for; he was in truth rather solemn — one might say ponderous and without much wit or humour. He did not appeal much to students, for he had none of the tricks and mannerisms which they like; he was, however, a sound teacher and a good all-round paediatrist. His special interest was in biochemistry — particularly rare metabolic disorders and he got excited when he came across a case of alkaptonuria or porphyrinuria ... He was always very kind to me when I first came to work with him as a young, raw, and (I fear) rather bumptious Scot and when I once remarked to him that London was the most provincial city in England (to which I still think there was some truth), he smiled tolerantly.[17]

A similar view was expressed by Robert S. Frew, who was associated with Garrod at Great Ormond Street and was also interested in metabolic diseases of childhood. A few years before his death, in a letter to C. J. R. Hart, Frew reminisced:

Fig. 17. Garrod in 1899, at the time of his appointment as Senior Physician at the Hospital for Sick Children.

He was the greatest scientific physician I encountered during my whole career, with a great gift for imparting his knowledge. His rounds lasted for about $1\frac{1}{2}$ hours, twice weekly, and half that time was occupied testing urines. He said to me once 'Clinical medicine is not really my main interest, I am a wanderer down the by-paths of medicine.'[18]

Garrod's alkaptonuric patients were his special interest.

He used to love bringing his alkaptonuric patients back time and again to the ward, and a case of steatorrhoea was heaven to him. He had both a male and female alkaptonuric and he used to get them into the ward at the same time, in the hope, as he said, that they might become fond of each other, for there was no known case of two alkaptonurics having married each other![19]

Frew remembered that 'The Laws of Mendel had only recently been "rediscovered" and Garrod would hold forth on Mendelism and illustrate it on many occasions.'[20] It seems likely that in doing so Garrod was reviewing how Mendelism had helped him understand his patients with inborn errors of metabolism, rather than displaying a fascination with Mendelism as a branch of science. Although there were a few physicians, like Frew, who appreciated Garrod's approach to medicine, it seems likely that there were many more who regarded Garrod as someone interested only in rare diseases that they would not encounter in their consulting practices.

When Garrod was invited by the medical students at Bart's to give a lecture to the Abernethian Society in 1899, he chose as his subject 'Some clinical aspects of children's disease.' With his customary charm and literary elegance, he argued that the time had come to regard the diseases of children as a special branch of medicine. The lecture was a general one, and covered many aspects of childhood. At every turn, however, his delight in children was apparent:

Children cannot fail to appeal strongly to your sympathies, ... whose diseases cannot be ascribed to any fault of their own, but are too frequently attributable to the ignorance and even neglect of their elders.[21]

Garrod's view that there was something special about diseases of children became widely known, and in 1904 the Board of Governors at St Bartholomew's created, at Garrod's urging, an out-patient department for children. Garrod was placed in charge, with Herbert Morley Fletcher as co-director.

It had been clear from the very beginning of Garrod's career that clinical practice alone, whether pediatric or adult, would never be enough to satisfy his scientific curiosity. As the years went by, he became increasingly convinced that the only way to understand human disease was to examine the chemical changes that might underlie or accompany a disease process. He had seen at first hand the usefulness of this approach in the case of gout, and he vividly recalled how his father had managed to obtain crystals of uric acid on cotton threads suspended in a watch-glass on the mantelpiece in his consulting room. When the test was performed on patients suffering from rheumatic diseases other than gout, uric acid crystals would not assemble on the threads. This test, popularly known as Garrod's thread test, came into wide diagnostic use.

Under the guidance and enthusiasm of Harcourt, chemistry had featured strongly in Archibald Garrod's studies as an undergraduate at Oxford. He knew intuitively that if chemical aberrations in disease were to be given serious attention, he would have to master the most sophisticated and precise scientific methods of chemical analysis. This was not a view shared by many of his clinical colleagues. Certainly his physician colleagues at the West London Hospital, as well as those at Bart's, were not particularly interested in applying the tools of science to disease. Although a few of them might have recognized, in a rather abstract sense, the need for a chemical approach, they did not have sufficient knowledge or interest to pursue further chemical studies.

As Garrod delved more deeply into the abnormal chemistry of disease processes, he, like his father before him, turned to an examination of the urine. Physicians had for centuries found this a profitable avenue of inquiry because, as Garrod put it, many diseases 'advertise themselves' through visible and chemical changes in the urine. Hippocrates, writing several hundred years before the birth of Christ, had said, 'Clouds carried about in the urine are good when white, but bad when black.'[22]

Garrod's first opportunity to investigate the chemistry of human disease in a detailed and systematic fashion came in 1892, when he encountered a patient with chorea who excreted urine that was tinted red. With considerable patience and using all his chemical and spectrographic skill, Garrod was able to show, in a detailed and scholarly article, that the color was caused by the presence of hematoporphyrin, and that this substance was also present in the urine of patients with 'articular rheumatism' and a variety of other diseases. The similarity of the urinary findings in patients with chorea and rheumatism supported Garrod's contention that the two diseases might have a common cause. However, after developing an effective method for separating hemato-porphyrin from similar urinary pigments, he concluded, evidently with some sorrow, that the pigment probably had nothing to do with chorea, for it occurred in normal urine also, although in very small quantities.

This disappointing result did not cause Garrod to despair; indeed, all his life he was to remain captivated by colored compounds. Red urine caused by porphyrins, yellow urine from urobilin, even green feces resulting, as he thought, from typhoid fever, all excited his attention, and to these variations he applied his analytical mind and chemical expertise. Throughout the 1890s, despite his increasing clinical responsibilities and teaching duties, Garrod continued to be fascinated with the chemical causes that led to the excretion of colored urine. His concurrent clinical appointments at three London hospitals afforded him ample opportunity to examine the urine of patients with many ailments. He particularly liked to emphasize the correlation of biochemical abnormalities with clinical disease. He was also always alert to the possibility that the presumed 'abnormal' constituent might be present in the urine of normal individuals, although in trace amounts. Garrod saw himself primarily as a clinician interested in the science of medicine, rather than as a scientist interested in clinical problems.

It was at about this time that a young analytical chemist at neighboring Guy's Hospital, Frederick Gowland Hopkins, entered Garrod's life (Fig. 18). Four years younger than Garrod, a future Nobel Laureate, a founder of British biochemistry, and a future president of the Royal Society, Hopkins shared many of Garrod's interests.[23] As a young man, Hopkins, like Garrod, had been a keen naturalist and lepidopterist, and in 1878, at the age of seventeen, he had published his first scientific paper, a short note in the *Entomologist* on the Bombardier beetle. In a paper published in 1889 Hopkins had outlined his belief that the yellowish-white pigments found in the

Fig. 18 Sir Frederick Gowland Hopkins, in a 1938 portrait. By kind permission of the President and Council of the Royal Society.

wing-scales of many butterflies, including the common cabbage white, were due to the presence of uric acid. Unlike Garrod, Hopkins had received training in forensic and analytical chemistry. He first entered Guy's as a chemical assistant to Sir Thomas Stevenson, a forensic analyst, but later decided that it would be beneficial to his career to possess a medical degree. He enrolled as a medical student at Guy's in 1888 at the age of twenty-seven, intending to become a lecturer in forensic medicine. Despite Hopkins's relative youth, clinicians at Guy's such as Frederick Pavy, whose specialty was metabolic diseases and diabetes, continually asked for his help in the metabolic investigation of their patients. In this way Hopkins obtained considerable experience in medical biochemistry; between 1891 and 1893 he accumulated enough material to publish three papers on the measurement of uric acid in the urine.

Whether Garrod had known of Hopkins before 1891 is not clear, but it seems hardly possible that both Garrods, father and son, fascinated as they were by uric acid metabolism, and being omnivorous readers of the scientific literature, had not encountered Hopkins's papers and read them with acute interest. In any event, Archibald Garrod immediately took a great liking to Hopkins, and they became close friends as well as effective scientific collaborators. Their friendship was to last all Garrod's life; and when Garrod died Hopkins wrote his obituary for the Royal Society.

Garrod and Hopkins turned their attention to an analysis of the pigments that give urine its yellow color, and correctly ascribed the color to the presence of urochrome, then known as uroerythrin. In 1896 Garrod focused on urobilin, and described much that was then new, including the spectroscopic properties of this pigment. He showed, for instance, that, contrary to common belief, urobilin obtained from urine, feces, and bile was identical. His two papers on urobilin, written with Hopkins in 1896 and 1897 and published in the *Journal of Physiology*, are models of chemical scholarship applied to human disease.[24]

Garrod's early papers on urohematoporphyrin were, for the most part, descriptive, and written with meticulous attention to chemical detail. In addition to extensive spectroscopic analysis, he undertook the laborious chemical isolation necessary to identify specific compounds. In 1886, Charles MacMunn and Ray Lankester, who had been at Oxford at the same time as Garrod, had brilliantly exploited the technique of spectroscopy to investigate the nature and structure of biological pigments. MacMunn first drew attention to the presence in the urine of a pigment he called urohematoporphyrin.[25] Garrod and Hopkins followed his lead and improved on his spectroscopic techniques.

Between 1892 and 1898 Garrod and Hopkins published 12 papers which dealt with the presence of urobilin and hematoporphyrin in the urine.[26] Although Hopkins's strong analytical background was undoubtedly helpful, Garrod's name appeared as the first author on the papers, and he probably performed the critical spectroscopical analyses. In 1898 Hopkins left Guy's for Cambridge; and, although he and Garrod remained close friends, there was no opportunity for them to collaborate in their further scientific research.

Garrod was not a scientific recluse. At all stages of his career, he took pains to show the relevance of chemistry to clinical medicine and to assure hesitant clinicians that they should not be overawed by science. In a general article entitled 'The spectrographic examination of the urine', published in the *Edinburgh Medical Journal*, he attempted to persuade clinicians that they should not be deterred from undertaking chemical analysis themselves.[27]

Porphyrins excited Garrod's chemical interests throughout his life. In modern terminology, hematoporphyrin is a 2-carboxylic porphyrin, which is not normally produced by any known physiological metabolic pathway. In rare instances, it has been reported that anaerobic infections of the bladder can lead to the presence of hematoporphyrin, or a hematoporphyrin-like molecule ascribed to the catabolic activity of bacteria. (This would be a remarkable metabolic feat, since the standard way of preparing hematoporphyrin in the laboratory is to boil hemoglobin in 6 normal hydrochloric acid.) It is therefore more likely that the urinary hematoporphyrin detected by Garrod was in fact coproporphyrin, a 4-carboxylic porphyrin. Observed in acid solutions, the spectrum of each compound is similar. Coproporphyrin is known to be excreted in the urine of normal individuals and in the urine of many of the conditions in which Garrod reported an increased excretion of urinary hematoporphyrin. Garrod showed that urinary hemato-porphyrin can be greatly increased in patients who take sulfonal, and thereby induce an acute attack of porphyria. There is a remote possibility that protoporphyrin may have accounted for some of the absorption patterns reported by Garrod. However, this porphyrin is not excreted in the urine.

Uroerythrin, a chromophore (color-causing chemical) which displays a pink color, occupied much of Garrod's attention. For a long time uroerythrin was thought to be yellow; however, the color was almost certainly due to the increased excretion of urobilin in liver disease and occasionally in thyroid disease. It is greatly to Garrod's credit that he stated definitively that uroerythrin is a distinct entity that bears no relationship to urobilin. In the ninety years or so since Garrod reported his studies on uroerythrin neither the chemical structure nor the biosynthetic pathway of the compound has been determined.

While he pursued his research with colored urine, Garrod continued clinical work and maintained his enthusiasm for recording his findings. Like most of his contemporaries, Garrod encouraged the young physicians who worked with him to write up their case reports and present them to the various clinical societies in London. It should be appreciated, however, that Garrod was unusual in approaching medicine from the chemical viewpoint. Most of the physicians in London specialized; usually this meant concentrating on diseases of the heart, as Sir James Mackenzie did. Sir Humphrey Rolleston was regarded as a liver expert, and his colleague Wilmot Herringham specialized in diseases of the kidney. Garrod did not take any organ as his own. He clearly recognized genetics as the basis of all biology; but medicine and its perturbations, as reflected in metabolic disease, were always his principal interest.

In 1893, Garrod described peculiar 'nodular excrescences' in the neighborhood of the finger joints.[28] Although Garrod suggested that these nodules might be features of a previously undescribed syndrome, it seems far more likely that the three patients represented a hetero-geneous group of nodular conditions. Painfully aware that no single preferred therapy was available for patients with chronic rheumatic disorders, Garrod agreed to visit a number of health spas in England and on the Continent to evaluate the existing regimens' usefulness. In 1891, while on holiday in Germany, he visited the spa at Schwalbach, about nine miles from Wiesbaden. After examining the evidence in some detail, he concluded that patients with chlorosis and iron-deficiency anemia indeed benefited from taking the waters. This was hardly surprising, since, as Garrod pointed out, the water at Schwalbach contained a large amount of iron. He also believed, though with much less conviction, that patients with 'uterine disorders', including 'a tendency to abortion and sterility', improved.[29]

Four years later, in the company of William M. Ord, a senior physician at Guy's Hospital, Garrod visited many of the medicinal springs of Great Britain as a member of a committee set up by the Royal Medical and Chirurgical Society.[30] Garrod and Ord visited and reported extensively on the water at Bath, Buxton, Matlock, and Droitwich. The reports were generally favorable, and one has to conclude that Garrod, like other physicians of his day, believed the waters offered some therapeutic benefit, not only to patients with muscular and articular disease, but also to those suffering from a variety of other disorders, including diseases of the liver. These findings are in startling contradiction to Garrod's hard-nosed search for scientific truth. Beset with patients suffering from various diseases for which there was no treatment, and perhaps influenced by his father's interest in materia medica, Garrod clutched at therapeutic straws. The reports that listed Garrod as first author, however, were always more conservative in their claims of therapeutic usefulness.

In 1896, when Sir Clifford Allbutt, Regius Professor of Physic at Cambridge, undertook the editorship of an eight-volume textbook, *A system of medicine*,[31] he turned to Garrod for the chapters on rheumatoid arthritis and other arthritic joint diseases. The book quickly became a classic, and went into many editions. Although Garrod did not contribute to all editions, his professional reputation, particularly in the field of rheumatic diseases, was greatly enhanced by this popular textbook.

While Archibald was busy seeing patients, publishing widely on all aspects of clinical medicine and pursuing his scientific interests, Laura gave birth to two more sons and a daughter. Their daughter, Dorothy Annie Elizabeth, was born on 5 May 1892, four and a half years after the birth of their son, Alfred Noël. In 1894, on 24 August, a second son, Thomas Martin, arrived; and on 29 December 1897, their last child, Basil Rahere, was born. Basil was named after the monk Rahere, who, returning from a pilgrimage to Rome, had a vision in which St Bartholomew told him that he must build a hospital in the city of London at Smithfield. Rahere obeyed, and in 1123 built St Bartholomew's Hospital.

Garrod seems to have had little appetite or time for the London social scene. His friends were mainly neighbors and colleagues (Fig. 19), and whatever free time he had he liked to spend with his growing family (Fig. 20). Dorothy recalls that from the age of nine:

His interest in astronomy, like that in history, always remained with him. Both these interests he imparted to his children at an early age, having a great gift for making such things interesting to the young... One of the great pleasures of the week was the Saturday afternoon walk, which was always the occasion for a long story, usually historical. In this way I learned the history of the Indian mutiny, and the chronicles of the Roman Emperors, which were developed in front of their busts in the British

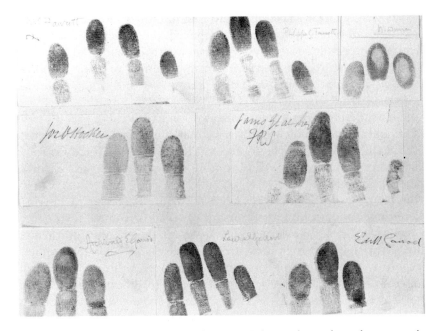

Fig. 19. On one visit to the home of Francis Galton, who undertook an extensive study of fingerprints, Archibald entered his prints into Galton's collection, as did his wife Laura and his sister Edith Garrod. Galton's array of prints reflected the fascination with notable families explored in his book *Hereditary genius;* along with the Garrods, this page includes the prints of noted scientists James Glaisher and Joseph Hooker.

Museum, and a host of other things. If possible my father planned the walk to take in some building or statue related to the story he told.[31]

The Garrods were living at their home at 10 Chandos Street, and their four children, even with considerable domestic help, kept Laura busy. The family had always been friendly with Gowland Hopkins and his wife, Jessie; but after 1898, when Hopkins went to Cambridge as Professor of Physiology, they saw each other less frequently, and usually only at scientific meetings. William Bateson, the champion of Mendelism, and his wife, Beatrice, were also living in Cambridge, and became good friends of the Garrods. The Batesons' firstborn, John, was the same age as Basil, and the boys undoubtedly played together whenever their families met. Both boys were to perish in the First World War.

Fig. 20. A Garrod family photograph, *circa* 1895.

(*L–R, back row:*) Effie Garrod; Archibald Garrod
(*middle row:*) Lucy Colchester Garrod, with Roland Garrod in arms; Herbert
Baring Garrod; Sir Alfred Garrod, with Dorothy Garrod on his lap; Edith
Garrod; Laura Garrod, with Thomas Martin Garrod
(*front row:*) Alfred Noël Garrod; Ralph Garrod; Geoffrey Garrod; Guy Garrod

Although Garrod served both the medical school and the hospital of
St Bartholomew in various capacities, it was not until 1903 that he
became, at the age of forty-five, an Assistant Physician to the Hospital.
So many capable physicians were senior to Garrod that, despite his
international reputation both as a clinician and as a scientist, he would
have to wait nine more years for a vacancy on the full consulting staff.

5

◆

Alkaptonuria:
The Clinical Clue

GARROD had been elected to the Royal Medical and Chirurgical Society as a young man, and it was to this society that he turned to present a paper on 9 May 1899. The paper, modestly entitled 'A contribution to the study of alkaptonuria', launched his epoch-making concept of inborn errors of metabolism.[1] It was later recognized as an intellectual leap forward that would fulfill all the criteria of Thomas Kuhn for a scientific revolution, a conceptual breakthrough whose consequences continue to have the greatest theoretical and practical significance for physicians, geneticists, and biochemists.[2]

Alkaptonuria, at the time known colloquially as 'black urine disease', was easily recognizable clinically because those affected passed urine that darkened when exposed to the air. Although extremely rare, alkaptonuria was not a new disease. In 1822 Alexander Marcet, a Physician at Guy's Hospital, presented to members of the same society the case of a seventeen-month-old infant whose urine had turned black soon after it was passed.[3]

The term was not used until 1859, when Carl H. D. Bödeker, Professor of Physiological Chemistry, encountered a patient presumed to have diabetes. The patient's urine reduced Fehling's solution, but did not reduce bismuth hydroxide, as it should have done if the reducing substance were glucose. When the urine was greatly concentrated, a small excess of glucose was present. But Bödeker, an astute observer, was struck by the fact that the urine darkened when exposed to alkali, and that it darkened from the surface.[4] For these reasons, he called the unknown reducing substance an 'alkapton' — a barbarous word, as Knox has pointed out, because it contains Greek and Arabic roots ($\kappa\acute{\alpha}\pi\tau\epsilon\iota\nu$ and *al-kali*) that are intended to denote the substance's rapid absorption of oxygen when exposed to alkali or room air.[5]

For several years after Bödeker's discovery, the exact nature of the black material remained obscure. Because alkaptonuric urine reduced

Fehling's solution, alkapton was often mistakenly identified as glucose. In 1891, the nature of the second reducing substance in the urine was clarified. Michael Wolkow and Carl Bauman were able to crystallize from the urine of alkaptonuric patients a compound, 2,5 dihydro-xyphenylacetic acid, which they termed 'homogentisinic acid' (soon to be renamed homogentisic acid).[6] An increase in the excretion of homogentisic acid followed ingestion of the amino acids tyrosine and phenylalanine. A third reducing substance had been reported to occur in the urine of a few patients with the disease; but Garrod insisted that the only constant urinary abnormality in alkaptonuria was the increased excretion of alkapton.

Wolkow and Baumann had suggested in their 1891 paper that the finding of alkapton in the urine of patients with alkaptonuria was probably due to the action of micro-organisms in the human intestine that converted tyrosine to homogentisic acid. As a result of his own investigations, Garrod seriously questioned this hypothesis. His increasing experience with alkaptonuria strongly suggested that the condition was lifelong, and seldom caused the patient any adverse effects. Occasionally, as Garrod was later to recognize, alkaptonuria may lead after many years to a harmless blackening of cartilage. Mild arthritis of the spine, though uncommon, was also very characteristic (Bödeker's patient developed disabling arthritis of the back later in life).[7] This syndrome, about which Garrod would subsequently write, was called ochronosis. The first case of ochronosis was observed by the great German pathologist Rudolph Virchow, who had reported blackening of the cartilages in a sixty-seven-year-old man with 'arthritis deformans'.[8] The disease was also of particular interest to William Osler of the Johns Hopkins Hospital in Baltimore, with whom Garrod had much correspondence.[9]

Garrod postulated that alkaptonuria was due to a chemical aberration, a freak, or as he was later to call it, an inborn error of metabolism. To pursue this possibility, he analyzed urine from patients suffering from alkaptonuria. In October 1898, Garrod published an article entitled 'Alkaptonuria: a simple method for the extraction of homogentisinic [*sic*] acid from the urine'.[10] In this short paper, Garrod focused entirely on the development of a method for the reliable chemical extraction of the compound. It is testimony to the importance that Garrod attached to the need for meticulous chemistry in the investigation of human disease that this purely methodological paper preceded the clinical paper.

Garrod's presentation to the Royal Medical and Chirurgical Society in 1899 was devoted largely to a discussion of the chemical abnormalities found in the urine.[11] But he also mentioned that frequently more than one family member was affected by the disease.

Following Garrod's presentation, Frederick W. Pavy, Physician at Guy's Hospital, rose to his feet to open the discussion. Four of the five patients that Garrod had discussed were under Pavy's care. Although now seventy years old, Pavy was still one of the most respected and distinguished scientifically minded physicians of his day. He, like Garrod, had a special interest in the chemical changes that occur in disease, particularly those that occur in diabetes. He was, of course, quite familiar with Garrod's findings:

As stated by the author, it so happens that four of the five cases of this rare condition had fallen under my observation. I had contemplated making use of them myself, but have been carried along by other work, and I was glad to find that Dr. Garrod was giving attention to this subject, and placed these cases unreservedly at his disposal, feeling sure that he would make better use of them than my time would allow me to do. His paper justifies my hopes, for he has really placed our knowledge of the subject on a sound basis. He mentions that it is seventy-seven years since the first case of the kind was recorded before the Society, but I scarcely think we shall have to wait as long for the next communication on the subject. It is a curious condition: it does not seem to produce any injury to health, but the urine, at first normal in appearance, gradually darkens and excites attention, especially in the case of infants, the staining of the napkins [nappies, diapers] exciting much domestic concern. Dr. Garrod has touched upon the possible cause of this condition. There is evidence against its being due to microbial activity. Many conditions of the urine are due to microbial life, as for instance indicanuria; but when we come to look not only at the evidence he has placed before us, but also at the fact that this condition has been noted from the earliest period of life, it is difficult to think that microbial life can be the cause. I suppose we must put it down to some erratic metabolism, to the chemistry of organic life not being carried on in a normal way; possibly one of the cogs of the machinery of life has been badly set, so as to lead to a wrong action, and so divert a process from its natural course. Although it does not appear to cause injury to health it may expose its subject to numerous inconveniences, either in the matter of life assurance or when he wishes to enter the services. Now that the examination of the urine is being carried out on such a large scale, it is possible we may come across this condition more frequently in the future than in the past. It occupies a somewhat analogous position; I do not mean in regard to its chemistry but in regard to other circumstances — to cystinuria. This too is due to some fault of metabolism, but with the latter we may have serious distress should the crystals not pass away but lead to the formation of calculi.[12]

Although Pavy and Garrod were in full agreement, the next speaker, Robert Maguire, was not. The same age as Garrod, Maguire was principally interested in renal diseases and tuberculosis. He was also regarded as a thoughtful and gifted physician, but he objected to Garrod's conclusions in the following terms:

I should like to point out that this, after all, is only one form of alkaptonuria, *i.e.* a condition in which the urine absorbs oxygen in the presence of alkali. Some fifteen years ago I called attention to the darkening of certain urines when exposed to the air after alkalinisation. That phenomenon is not peculiar to this particular homo-gentisinic acid. For instance, in carboluria, a condition which is not altogether innocent, it may be associated with poisons probably due to some action of the carbolic acid on the system.[13]

Later Maguire made another point:

I hold, therefore, that this term must not be limited absolutely to this particular innocent group of cases. Has the author ascertained the actual constitution of the pigment? I understand that it is a combination of this acid with oxygen in presence of an alkali.[14]

Gowland Hopkins then rose to defend his friend and collaborator; but he did not address the principal issue. Instead, he chose to empha-size the rarity of true alkaptonuria, as opposed to the occasional excre-tion of dark urine:

It was my fate some time ago to have to examine a series of over 3000 specimens of morbid urine, and of that whole series only one specimen showed at all unequivocally the classic signs of alkaptonuria. Out of the 3000 samples there were two others in which a vague darkening of the urine took place on exposure to the air. These figures are perhaps of more significance in that a very large proportion were submitted for examination on account of alleged abnormalities of pigmentation, or because they exhibited some vague reducing power. My material was therefore rather concentrated from this point of view. It seems to me very fortunate that these cases should have been preserved by Dr. Pavy and should have fallen into such able hands.[15]

In closing the discussion, Garrod, not in the least intimidated by Maguire's comments, held fast to his position. He readily admitted that he had not yet had an opportunity to investigate the nature of the pigment:

I do not think anyone has yet specially investigated the pigment. It is one of the group of substances which are formed by the oxidation of aromatic bodies, with which those of us who happen to be photographers are only too familiar. As regards pigments, we studiously investigate those of which we do not know the precursors. Urobilin, for example, is one of which much attention has been given, but if we could isolate the chromogen from which it is derived, it would probably prove to be a still more interesting object of study. I do not agree that the term alkaptonuria should be extended to the cases alluded to by Dr. Maguire. It is stated in books that urine containing gallic acid gives a very similar action, and I recently had the opportunity of examining some specimens. It did darken with alkalies, but to nothing like the same extent, and the reducing power was quite insignificant. I think it would be well to keep this term to this particular group of cases, whilst acknowledging their kinship to other cases in which urine darkens on standing as a result of the oxidation of aromatic substances.[16]

Garrod had described five new cases of the disease in his initial presentation, and he was determined to collect and analyze all the cases of alkaptonuria in the literature. In the winter of 1901 he wrote to Osler in Baltimore to enquire about the frequency of alkaptonuria in the United States.[17] He also corresponded with physicians in Germany and Italy, who were eager to provide as much information as they could. As Garrod learned more about the disease, he found an increase in consanguinity among the parents of those affected, and he became more and more convinced that alkaptonuria was the result of an inborn chemical error, or a 'metabolic sport', as he then called it.

Intrigued by the possibility that this congenital disease could be inherited, Garrod was disappointed not to be able to find, either in the medical literature or among his own patients, an instance in which the disease was transmitted from one generation to the next, as had been shown in a number of other inherited disorders. It would be of considerable interest to know when Garrod first hit on the idea that alkaptonuria was a metabolic disease. Although the evidence is anecdotal, it seems possible that the idea came to him suddenly. C. J. R. Hart gives the following account, from recollections of George Graham:

Garrod was never totally satisfied by the condition being due to an infection and it was "one afternoon while walking home from hospital and pondering on this problem, it flashed through Garrod's mind that the condition was perhaps caused by a metabolic error, a disturbance of chemical activity in which a different molecule resulted as an end-product. He saw that the origin of such a state of affairs could be hereditary. Fortunately, the mother of one of his alkaptonuric patients was pregnant at the time, and Garrod resolved to test it."[18]

In November 1901 Garrod read to the Society a second paper, 'About alkaptonuria' an abstract of which was immediately published in the *Lancet*,[19] and subsequently in the *Medico-Chirurgical Transactions*.[20] Garrod began:

The object of the present communication is to call attention to certain facts and to record some observations which tend to throw fresh light upon its nature and causation.

He drew specific attention to the frequency of first-cousin marriages among patients with the disease:

That alkaptonuria may be met with in several members of a family was first pointed out by Kirk in 1886, [21] and of the cases since recorded a considerable number have served to emphasize this fact. However, although brothers and sisters share this peculiarity, there is, as yet, no known instance of its transmission from one generation to another, nor is anything known as to the urine of children of alkaptonuric individuals.

On the other hand I am able to bring forward evidence which seems to point, in no uncertain manner, to a very special liability of alkaptonuria to occur in the children of first cousins. The information available relates to four families including no less than eleven alkaptonuric members, or more than a quarter of the recorded examples of the condition.

Later in the paper he asserted:

The facts here brought forward lend support to the view that alkaptonuria is what may be described as a "freak" of metabolism, a chemical abnormality more or less analogous to structural malformation. They can hardly be reconciled with the theory that it results from a special form of infection of the alimentary canal. There is here no question of the intensification of family tendencies by intermarriage, for in no instance were the parents themselves alkaptonuric, and, as has been already mentioned, there is, up to now, no recorded instance of alkaptonuria in two generations of a family.

Garrod wanted to provide unequivocal evidence that alkaptonuria was congenital, in the sense that the condition was present from birth. He followed all his patients with alkaptonuria closely, and he knew that the mother of one of them was pregnant with her fifth child. When the baby boy was born on 1 March 1901, Garrod carefully instructed the nurses to examine the napkins (diapers) for any trace of darkening or staining. None had appeared 15 hours after birth. By the morning of the second day, however, the napkins were slightly stained, and by 10.30 a.m. on 3 March, just 52 hours after birth, the infant's napkins were stained black. This crucial evidence enabled Garrod to state with assurance that alkaptonuria was a congenital disorder, probably attributable to an error in the chemical make-up of the body:

The above facts, carefully recorded by one who was wholly without bias in favour of any theory of the nature of alkaptonuria, or knowledge of the questions at issue, nevertheless agree completely with what was to be expected on theoretical grounds.[22]

He concluded by addressing a more subtle point — whether the increase in homogentisic acid that follows a protein meal is merely due to a chemical transformation from tyrosine to homogentisic acid in the intestine, as Franz Mittelbach believed,[23] or whether, as Garrod firmly believed, the defect in alkaptonuria was in the tissues. By feeding one of his patients a high-protein meal, he was able to show that the increase in homogentisic acid in the urine occurred at its maximum four to six hours after the meal, not two to three hours after, as Mittelbach had claimed:

In a word, they tend to support the view that the change from tyrosine to homogentisic acid takes place in the tissues after the absorption of the former, rather than the alternative view that the change in question is brought about in the alimentary canal.

In the discussion following Garrod's presentation, W. A. Osbourne mentioned the case of a man who was rejected for life assurance because his urine reduced Fehling's solution due to alkapton. It might be a good plan, Osbourne continued, 'to give an alkaptonuric patient some of the intermediate substances as between tyrosine and homogentisic acid, and observe the effect on the excretion of alkapton in the urine.' Garrod responded by saying:

It would be difficult to give a tyrosin-free diet in his case, as the patient was a child of four years. The experiment had been tried abroad by Mittelbach, whose adult patient had consented to take only tea and brandy for three days. Mittelbach found that after such fasting the homogentisic acid excretion fell to about one third of the usual amount, but that the acid did not completely disappear from the urine.[24]

Despite Garrod's emphasis on consanguinity, the possible hereditary nature of the disease was not made explicit in this paper.

In 1900, while Garrod's studies on alkaptonuria were progressing, Gregor Mendel's work on heredity was being 'rediscovered' by Hugo de Vries in Amsterdam, Carl Correns in Tübingen, and Erich Tschermak in Vienna.[25] Not surprisingly, Mendel's startling results encouraged scientists to look at their biological experiments from an entirely new perspective. William Bateson, a botanist, was one of Mendel's keenest and most aggressively articulate disciples. In 1909, he published *Mendel's principles of heredity,* and in this classic explained the hereditary principles that he believed Mendel had uncovered in his study of garden peas.[26]

Garrod's discovery that first-cousin marriages tended to yield a higher percentage of alkaptonurics was, of course, highly significant. In 1901, however, he was not sufficiently aware of Mendelian laws to understand why the marriage of first cousins would be more likely to produce an alkaptonuric child. Not until he discussed the matter with Bateson did he begin to appreciate the importance of consanguinity in understanding hereditary diseases.

On learning of Garrod's work, Bateson quickly realized that alkaptonuria might be an example of a recessive hereditary trait. On 17 December 1901, he reported to the Evolution Committee of the Royal Society the genetic significance of Garrod's observation that the parents of patients with alkaptonuria were frequently first cousins:

In illustration of such a phenomenon [the persistence of the hidden recessive character] we may perhaps venture to refer to the extraordinarily interesting evidence lately collected by Garrod regarding the rare condition known as "Alkaptonuria." In such persons the substance, alkapton, forms a regular constituent of the urine, giving it a deep brown colour which becomes black on exposure.

The condition is extremely rare, and, though met with in several members of the same families, has only once been known to be directly transmitted from patient to offspring. Recently, however, Garrod has noticed that no fewer than five families containing alkaptonuric members, more than a quarter of the recorded cases, are the offspring of unions of *first cousins*. In only *two* other families is the parentage known, one of these being the case in which the father was alkaptonuric. In the other case the parents were *not* related.

Now there may be other accounts possible, but we note that the mating of first cousins gives exactly the conditions most likely to enable a rare and usually recessive character to show itself. If the bearer of such a gamete mates with individuals not bearing it, the character would hardly ever be seen; but first cousins will frequently be bearers of *similar* gametes, which may in such unions meet each other, and thus lead to the manifestation of the peculiar recessive characters in the zygote.[27]

In late 1901 Bateson and Garrod had begun a rapid-fire correspondence in which they discussed the probable significance of consanguinity in relation to alkaptonuria. It is difficult to know whether Garrod first sought Bateson's assistance, or if Bateson wrote to Garrod because he had heard about his work.

In a long letter dated 11 January 1902, Garrod begins rather formally:[28] 'Dear Sir, It was a great pleasure to receive your letter and to learn that you are interested in the family occurrence of alkaptonuria.' This reference to an earlier letter might indicate that Bateson initiated the correspondence after having read or been referred to Garrod's article in the *Lancet* in November 1901. Garrod goes on to discuss the known incidence of alkaptonuria throughout the world, and indicates he is aware of only 37 recorded cases. He then poses a different question: 'Do you attach any importance to the kind of first cousinship? In the only cases in which I have such information the parents were the *children of sisters*.'[29]

There is no record of Bateson's replying to Garrod's question, and Garrod does not raise it again. He was almost certainly familiar by this time with the contents of the *Report to the Evolution Committee of the Royal Society*,[30] of which Bateson was secretary. Garrod seems to have realized that consanguinity was a crucial factor in assessing the mode of inheritance in alkaptonuria, and although he did not use the language of genetics, it seems that he understood the concept of recessive inheritance, as is clear from his letter to Bateson:

I am convinced that alkaptonuria is actually and not merely apparently rare — and this renders it very improbable that two *actually* alkaptonuric strains should unite, but may it not be that the condition results from the union of two *potentially* alkaptonuric strains, and that such potentiality is much less rare — I am afraid that for the same reason a marriage of alkaptonurics is very unlikely to occur, nor do I see my way of introducing any marriageable alkaptonurics to each other with a view of matrimony!!

Any further suggestions as to lines of investigation I should *greatly value, and would try* to follow out as far as possible. The subject interests me greatly in its bearing upon chemical as distinguished from structural variations, and it seems to me that alkaptonuria, cystinuria and perhaps albinism also are chemical analogues of malformations. I should be very glad indeed to know your opinion upon these ideas, and at the same time, I would ask you kindly not to speak of them to others, insofar as they may contain anything new. It may be that they are quite familiar to those who, like yourself, have devoted much thought to these subjects, and if so I should be very glad to know it. At any rate a collection of evidence as to specific generic and individual chemical differences might prove of some value.

Please excuse the length of this letter and believe me,

Yours truly,
Archibald E. Garrod.[31]

Later in the same month, in a long and friendly letter to Bateson, Garrod discussed his views on malfunctions, but again came back to the important question of consanguinity in recessively inherited diseases:

I am much encouraged by what you say about chemical variations, and shall hope in the course of a few years to get together some useful information on the subject.

I fancy that monstrosities or rather malformations, vestigial remnants, and individual differences, all have their chemical analogues, but I cannot think of any animal which is domesticated and bred for the sake of a chemical product, although among plants examples are common enough.

In the case of cystinuria the state of affairs is different from that in alkaptonuria, as regards heredity. It also occurs in families, and there are good examples known of its transmission by *either* parent. In the only two cases which I can get hold of there is no relationship of parents. There are rather more than 100 recorded cases. Hopkins is an old friend of mine and we have written several papers together. His butterfly work has always struck me as extremely suggestive.

I am having enquiries made about every child admitted to my ward at the Children's Hospital, and purpose looking for other peculiarities of the urine of children of first cousins.

Both alkaptonuria and cystinuria are conditions which *advertise* their presence, and it seems probable that there are other freaks of metabolism which do not do so. If these are equally rare the task of looking for them in urine generally seems almost hopeless, but if one limits oneself to the children of cousins the field will be very much narrowed.[32]

Garrod was the first physician to appreciate the significance of Mendelism for human diseases and the special importance of consanguinity in those diseases that are recessively inherited. One year after Mendel's laws had been rediscovered, Garrod was confidently relating genetic concepts to chemical variations in man:

I have for some time been collecting information as to specific and individual differences of metabolism, which seems to me to be a little explored but promising field in relation to natural selection, and I believe that no two individuals are exactly alike chemically any more than structurally.[33]

By 1902 Garrod had grasped the idea of human biochemical individuality and understood the biological significance of consanguinity in alkaptonuria. It had always been obvious that individuals looked different physically, but it was an entirely new and far-reaching concept to suppose that each person's chemical make-up was individually distinct. The concept of human biochemical individuality had been born.

On 20 March 1902, in another letter to Bateson, Garrod again tried to account for the rare reported occurrence of two generations of alkaptonuria in certain families. Although clearly puzzled, he was dogged in his search for more convincing evidence of vertical transmission:

I am gradually getting in some further information about alkaptonuria and heredity. I think I told you that the parents of Erich Meyer's patient are first cousins. Prof. Osler is kindly following up some American cases for me. He has not yet been able to find out about the parentage of his own patient, but has seen two of his sons, one of whom proves to be alkaptonuric and the other not. This is the first known instance of transmission from parent to child.

The metabolic error in these cases is evidently a definite and constant one. May we not look upon it as separated from the normal metabolism by a dividing line, and that the members of certain families run much closer to this line than do others. When members of such families intermarry, a certain proportion of their offspring overstep the line by intensification of the family tendencies.[34]

In these last sentences it seems that Garrod was musing on the notion that perhaps there are quantitative variations as well as absolute presence or absence. While it may be too extravagant to think that the germ of biochemical susceptibility was beginning to take shape, it is clear that Garrod's fertile mind was exploring the possibility of variation in expression of a trait between families.

Garrod continued to barrage Bateson with letters and in June 1902 wrote:

Thank you so very much for so kindly sending me the report to the Evolution Committee. Your note on alkaptonuria interests me very much indeed. The same explanation will of course apply to albinism which also seems to occur with considerable frequency in the children of first cousins, as in the remarkable instance quoted by Darwin in "Animals and Plants". My enquiries are getting on slowly. In one of the recent German cases (Ewald Sher) the parents are neither alkaptonuric nor related. I am hoping to hear in the course of a few days from Embden who has been making enquiries in a remote valley of the Black Forest about the cases described by Baumann and himself. Two alkaptonurics were born out of wedlock, then both parents married anyway and had families, none of whom were alkaptonuric. If it should prove that they were cousins it will indeed be interesting. I have great hopes of learning also about an Italian case which can still be traced, about a family with 3 alkaptonuric members in Germany, and about the family with direct inheritance in America. If so there will be eleven families available containing 23 of the 39 recorded alkaptonurics.[35]

For the next six months, Garrod combed the medical and scientific literature with renewed intensity for additional cases of alkaptonuria, with particular attention to the possibility of strong hereditary influences. In December 1902 he wrote about the incidence of alkapto-nuria, and this time was emboldened to have a subtitle, 'The incidence of alkaptonuria: a study in chemical individuality'. In this paper, he discussed 40 cases of alkaptonuria known to him (he also mentioned albinism and cystinuria as examples of possible inborn errors), and offered an explanation for the occurrence of the anomaly:

There is no reason to suppose that mere consanguinity of parents can originate such a condition as alkaptonuria in their offspring, and we must rather seek an explanation in some peculiarity of the parents, which may remain latent for generations, but which has the best chance of asserting itself in the offspring of the union of two members of a family in which it is transmitted. This applies equally to other examples of that peculiar form of heredity which has long been a puzzle to investigators of such subjects, which results in the appearance in several collateral members of a family of a peculiarity which has not been manifested at least in recent preceding generations. It has recently been pointed out by Bateson [*Mendel's principles of heredity*] that the law of heredity discovered by Mendel offers a reasonable account of such phenomena.[36]

In addition to recognizing that alkaptonuria is a chemical anomaly, Garrod suggested that other diseases, including in some instances con-genital malformations, could be traced to a hereditary chemical imbalance of metabolism.

The evolution of Garrod's thought is clearly evident in these three papers.[37] His first paper described the occurrence of the disease among siblings. The second emphasized consanguinity, but heredity is not mentioned. In the last of the three, alkaptonuria's inherited character had become clear to Garrod. A simple recessive form of inheritance was all that was needed to explain the virtual absence of vertical transmission of the disease, as well as the increased parental consanguinity. The story was complete.

6

◆

A Career Caught in Controversy

IN November 1900 Garrod was invited by Sir William Church, President of the Royal College of Physicians and a colleague of his at Bart's, to deliver the Bradshaw Lecture, established in 1880 in memory of William Wood Bradshaw. Although at the time the complexities of alkaptonuria were uppermost in Garrod's mind, the invitation was scientifically opportune. Over the previous decade, the chemistry of urinary pigments had been one of Garrod's great interests, and the invitation gave him an opportunity to organize his thoughts on the subject and present them to his fellow physicians. His paper was characteristically scholarly, revealing his familiarity with relevant Continental literature. In order to simplify his topic, Garrod decided to limit his discussion to urobilin, hematoporphyrin, urochrome, and uroerythrin. The lecture was dauntingly chemical, although Garrod made a special effort to discuss the pathological significance of an increase in the excretion of colored urinary compounds. It is difficult to believe that the audience, composed almost exclusively of practising physicians, was spellbound by the discourse. Garrod adverted to the clinical relevance of his work, but was careful not to overstate its usefulness for the general physician:

In this review of a somewhat obscure branch of chemical pathology nothing has been further from my desire than to claim for my subject an exaggerated practical importance or to suggest that the investigation of the urinary pigments should form part of the routine of clinical examination. Such importance as the subject possesses is derived from the light which it throws upon processes which are at work in the body both in health and in disease, and especially upon those which are concerned with the disposal of effete blood pigment. Nevertheless, in certain cases the examination of the colouring matters of the urine may afford information of real diagnostic value and which is not readily obtained in other ways. It may therefore be fairly claimed that their study is one which is not simply and solely of academic interest.[1]

It was also in 1900 that Garrod elaborated his ideas on cystinuria, the second inborn error of metabolism he described, and the one that allowed him to generalize the concept of inborn errors of metabolism to other metabolic disorders. He believed that the increased urinary excretion of cystine in patients with cystinuria probably reflected a disturbance of cystine metabolism, but he was puzzled by the intermittent appearance of small amounts of cadaverine and putrescine in the urine in these patients.[2] Garrod was relentless in tracking down families with metabolic disease around the world. He continued to write to William Osler regarding additional cases of alkaptonuria and ochronosis in the United States. Their letters were cordial, even after Osler magisterially requested that 100 reprints of Garrod's 1901 paper, 'About alkaptonuria', should be dispatched to him in Baltimore.[3] Garrod also continued his correspondence with Bateson, who had sharply rebutted Walter Weldon's attack on Mendelism. Garrod sent Bateson a pedigree of a family in which parents who were first cousins had two children with six fingers; he clearly suspected recessive inheritance, and hoped for Bateson's concurrence.[4]

Unlike many of his colleagues, Garrod was an unusually good linguist. French and particularly German came easily to him; he also had a smattering of Italian. In 1903, he contributed to the German scientific literature a paper on chemical individuality, by then a canonical principle in his scientific approach to clinical medicine. Unremittingly punctilious in matters of priority, Garrod gave credit to Carl H. Huppert of Prague, who in his remarkable 1895 rectorial address at the Carl Ferdinand University had stimulated Garrod to think more deeply about the evolutionary background of comparative physiology and chemical individuality. Huppert argued:

As carriers of chemical individuality, protein substances [are] accorded priority; they are the core to which all other elements of the organization subordinate themselves. A particular chemical composition of at least some, if not all, protein substances must be presumed for every animal species. The next question, then, will be whether the number of proteins suffices already and, if that is not the case, whether it must be anticipated that one day it will suffice in order to satisfy the need for the many individual types.[5]

Huppert believed that chemical differences must exist among species. Pointing to the well-known fact that animals and man respond differently to morphine, Huppert suggested that inter-species differences in susceptibility to infectious agents were due to chemical individuality. The mechanism that conferred this individuality rested, he believed, on the structural differences of proteins of the body. Huppert's extraordinary insights are seldom, if ever, alluded to. They were not lost on Garrod, however:

[The] various families, genera, and species of animals and plants differ just as much from each other in their chemical structure and the products of their metabolism as in their anatomical structure and form.[6]

In his 1903 paper Garrod analyzed for his German readers the 40 cases of alkaptonuria that he had published, and added cystinuria and albinism to the growing list of errors of metabolism akin to alkaptonuria. Although he had heard of one case in which alkaptonuria was reported to have occurred in two generations, he did not modify his view that the disease was inherited in a recessive fashion. He understood clearly that in an isolated population, where there is an increased frequency of consanguinity, the possibility that a patient with alkaptonuria might marry an individual who was a carrier of the gene for alkaptonuria was greatly increased. If such a union were to take place, vertical transmission might be expected to occur. (On the average, 50 per cent of the children born from such a union would have overt alkaptonuria.) Garrod believed that both albinism and cystinuria should be regarded as inherited 'chemical malformations'. He had learned the principles of Mendelism from Bateson, who referred to Garrod in his book *Mendel's principles of heredity*, published in 1909.[7]

Garrod had begun his chemical research in the laboratories at Bart's under the general administrative supervision of F. D. Chattaway, a distinguished organic chemist whose synthesis of the chloramines introduced one of the first effective disinfectants into medical practice. In 1900 Chattaway resigned his position to take up a post in Heidelberg. W. H. Hurtley, who had been at the school since 1899 and knew Garrod well, became Chattaway's successor. This proved to be a long, happy, and scientifically fruitful appointment for both Hurtley and Garrod. Garrod persuaded Hurtley to shift his research interest to the many still-unsolved chemical problems of alkaptonuria, as well as to other inborn errors of metabolism that had attracted Garrod's clinical interest. Between 1905 and 1908, and aided by T. Shirley Hele, they wrote several articles for *The Journal of Physiology* dealing with chemical complexities surrounding alkaptonuria and cystinuria.[8–11]

It had long been held that, in some patients with alkaptonuria, a second alkapton acid, called uroleucic acid, was excreted in the urine. If this were the case, some of Garrod's ideas might have to be modified. Hurtley, at Garrod's urging, undertook detailed chemical studies to evaluate the claim. The results showed clearly that uroleucic acid was not excreted in the urine of alkaptonurics. It is typical of Garrod's kindly disposition that he wrote understandingly about the error and the several coincidences that had led the authors to their conclusion.[12]

In January 1903, at the age of forty-five, Garrod 'begged [his Royal Highness, the President, the Treasurer and Governors of St Bartholomew's Hospital] to offer myself as a candidate for the post of Assistant Physician to your Hospital'.[13] The application must have been one of the most meritorious ever entertained by their solemn Lordships. Garrod was one of the most distinguished physicians in the country, but his promotion had been long delayed. He had unsuccessfully applied for the position in 1887, again in 1893, and twice in 1895. With the promotion of Norman Moore to the rank of Physician in 1903, another vacancy now arose. Garrod, however, was taking no chances. He submitted 29 testimonials to support his application, including one from his father-in-law, Sir Thomas Smith. The letters could not have been more enthusiastic, and their tone attested to the writers' embarrassment that circumstances had not allowed Garrod to be elected to the staff sooner. Garrod was known to his colleagues as a physician of the first rank, erudite and courteous. He was well known on the continent, particularly in Germany, and was a prolific contributor to the clinical and scientific literature, 'a man of wide general education, of thorough medical culture, and of prolonged and fruitful scientific research'.[14] None of the testimonials made any reference to his work on alkaptonuria or his interest in metabolic disease. It could be argued, with some validity, that Garrod himself was only just beginning to realize the profound ramifications of his research.

In October 1903 Garrod, emboldened by his new position as Assistant Physician, asked and received permission of the Dean of the School to deliver an introductory lecture on chemical pathology to the medical students at the hospital. In delivering this lecture, Garrod introduced a new subject into the medical school curriculum. Garrod gave ample credit to all those who had gone before, and only briefly referred to his own contributions. In describing to his medical school audience his vision of the changes that could be wrought by thinking of disease in chemical as well as pathological terms, Garrod returned to astronomy for an analogy:

It would be easy to multiply parallel instances to the present state of our knowledge of these newly explored regions of chemical pathology. There was a time when the planet Neptune was only known as an undiscovered member of the solar system which exercised a disturbing influence upon the orbit of Uranus; and much more recent examples are afforded by the discovery of argon, which was first detected as something which introduced an error in the determination of the density of nitrogen obtained from different sources; and that of helium, which had been recognized for a considerable time as yielding a line in the solar spectrum which could not be assigned to any known element.

During the previous ten years, Garrod had shown that chemical examination of the urine could yield a granary of information about man and his diseases, and he wished to impress this fact on his students:

The importance of the study of the urine as a branch of chemical pathology can hardly be over-estimated. The labour expended upon it has been immense, and no better evidence of the mass of knowledge acquired can possibly be afforded than by a glance through the monumental work of Neubauer and Vogel, now edited by Professor Huppert, the several editions of which accurately reflect the progress of urinary chemistry.[15]

Garrod received the new title of Lecturer in Chemical Pathology, and continued to lecture on the subject until 1919, when he left Bart's for Oxford.

Garrod was also invited in 1903 to write a paper for the *Practitioner*, a review journal designed to help general practitioners keep up to date. He could not resist writing about alkaptonuria; but, with a view to his audience, he gave his paper the more arresting title 'On black urine'.[16] In the paper, Garrod compiled a comprehensive list of conditions in which the urine, while not inky black, was distinctly darker than normal, and he reviewed the increasing evidence that long-standing alkaptonuria could lead to ochronosis. He correctly concluded that alkaptonuria is not present in all cases of ochronosis, and that blackening of the tissues may be due to other causes. Garrod could not refrain from discussing one of his favorite subjects, the value of spectroscopy as a diagnostic tool for investigating colored compounds — even though the technique would hardly have been generally available to the clinicians who were his readers.

Although Garrod had been associated with the Hospital for Sick Children for ten years, it was not until he was appointed Physician to the Hospital that children came directly under his care. He became particularly interested in young patients with Sydenham's chorea, a disease known colloquially as Saint Vitus's Dance. Characterized by incapacitating, generalized, involuntary movements of the face and body, its possible relationship to rheumatic disease was of considerable interest. Garrod saw many children with the disease, for it was common among the poor who sought medical attention at the hospital. In 1905, he gave a clinical lecture on Sydenham's chorea at the hospital and presented four patients with the disease, carefully chosen to demonstrate the variability in the severity of the choreiform movements. He discussed the importance of a meticulous clinical examination of such patients for other signs suggestive of rheumatic disease. Pointing out that children with the disease are often nervous and excitable, Garrod proposed that he and the students 'adjourn to the museum and there discuss our subject more fully', in order to avoid disturbing his patients unnecessarily.[17]

Seventy-five years later, Garrod's summary of the disease can scarcely be faulted. He believed that the association of chorea with rheumatic disease was beyond dispute. He approved of the use of salicylates, but found them often less effective in patients with chorea than in patients with overt rheumatic fever. He believed that the evidence of the benefits of arsenic was quite unpersuasive, and insisted that prolonged bed rest decreased the probability of permanent cardiac damage.

Throughout Garrod's varied professional life, whether looking after children or adults, much of his thinking was dominated by chemical pathology. His definition of chemical pathology and his view of chemical individuality are best expressed in his introductory lecture to a course on chemical pathology at Bart's in 1903:

Chemical pathology is only beginning to take a distinct place among the subjects which are taught in our medical schools ... Let me point out at the commencement that there are grounds for the belief that no two human beings are completely alike as regards their chemical processes, any more than they are identical in bodily form. That the species and genera of animals and plants exhibit more obvious chemical diversities is a fact which admits of no question. Examples will at once occur to you even among the higher animals, such as the differences in the composition and crystalline forms of the haemoglobins of different species and the varieties of the bile acids. Furthermore, there are reasons for thinking that in rare instances congenital deviations from the specific type of metabolism are to be regarded more as instances of extreme variation than as results of morbid processes rightly so called.[18]

The lectures contained remarkable insights. Garrod pointed out that urine and serous fluids, because they can be obtained in relatively large amounts, enable researchers to identify substances, such as intermediate metabolic products, that normally occur in concentrations too low to be detected in the blood. He insisted that if studies on the general disturbances of metabolic balance that occur in some diseases are to be really useful, the patients must have a strictly controlled diet and must be placed in a metabolic chamber that permits meticulous control of diet and metabolic rate. Such investigations, he said, 'afford the most complex means available for determining what is happening in the body laboratory. I refer to what are known as "metabolism observations".'[19]

Chemical pathology was becoming increasingly familiar to general physicians, as well as to hospital specialists, and this was in large part due to the influence of Gowland Hopkins and Garrod. However, Garrod's deep interest and skill in biochemistry and clinical pathology led physicians to regard him as being a little too chemical in outlook. Garrod's conviction that the discipline of chemical pathology would yield profound insights into human disease was by no means universally held.

While Garrod's interest in chemistry as a means of understanding medical problems was always to the fore, he was less fascinated with the emerging science of genetics. For Garrod's chemical mind, the concept of a gene (the term coined by Wilhelm Johannsen in 1909) was too vague and elusive to be of great interest. There was also the danger of becoming caught up in the heated but fruitless debates that plagued the decade.

In the first decade of the twentieth century, when Mendel's work became widely known, Bateson and, later, R. C. Punnett were among the group of biologists who first realized its importance for understanding human disease. Although principally regarded as a poultry geneticist, Punnett was interested in the general applicability of the laws of Mendel to members of the plant and animal kingdoms, including man.

In February 1908, a meeting was held by the Epidemiology Section of the Royal Society of Medicine. In a lecture to this group, perhaps the first of its kind to a general medical audience, Punnett outlined the principles of dominant and recessive inheritance, using as an example the variation in color of the Blue Andalusian Fowl, and described the general ability of Mendelism to explain variations in the size and coloring of plants.[20] Turning to human disease, he described pedigrees of night blindness and brachydactyly, which were consistent with dominant inheritance, and hemophilia, which was clearly sex-limited. Punnett urged his audience to collect human pedigrees. He pointed out that 'the evidence recently collected by Dr. Garrod on alkaptonuria points strongly to this condition being recessive to the normal' believing, with Bateson, that alkaptonuria was an example of 'the presence and absence hypothesis'. With very considerable insight, Punnett urged the chemists to isolate the hypothetical ferment that was presumed absent in alkaptonuria, for 'it would serve to clear up our ideas upon the condition known as the diathesis to a disease, and would offer the hope of these conditions falling within the scope of heredity and consequently becoming amenable to human control'.

Punnett divided diseases into three classes, of which the first two would be expected to exhibit Mendelian principles of heredity:

(1)   Diseases which depend directly upon a structural change in the gamete, either by the addition or subtraction of some character as compared with the normal, e.g., night-blindness, brachydactyly, alkaptonuria.

(2)   Diseases in which such structural change of the gamete is without visible effect, but which renders the individual liable to invasion by bacteria, &c. The disease is not manifested unless the structural change and the external organism are both present; e.g., rust in wheat.

(3)   Diseases caused by external invasion, for which immunity, as implied by gametic structure, is not known to exist, e.g., syphilis, ankylostomiasis.[21]

The audience contained several skeptics. H. M. Vernon took the ancestrian view that all heredity was based on the concept of quantitative continuing variation. While not able to attend the lecture, Vernon wrote a testy letter that was read to the group:

The vast amount of work done by Galton, Pearson and others on the transmission of such blended characters and their relation to the characters of the parents, grandparents, &c., was practically ignored by the Mendelians. For the average medical man a knowledge of the laws of ancestral heredity, as defined by the workers mentioned, appeared more important than a knowledge of the segregated transmission of a few very rare diseases, interesting as such cases were.[22]

Garrod, who was in the audience, spoke up in favor of:

the suggestion contained in Mr. Punnett's paper that alkaptonuria might result from the absence of an enzyme which brought about the disintegration of the benzene ring of the aromatic fractions of proteins, [and] said that this view had been suggested in several quarters on quite other grounds than those of heredity, and that from the standpoint of chemical physiology there was much to be said in its favour. He [Garrod] called attention to the difficulty of obtaining satisfactory evidence of the occurrence of such chemical "sports" in the families of the patients. Although alkaptonuria was a fairly evident anomaly it was not easy to find out whether members of back generations of a family had stained their napkins in infancy or had passed urine which darkened on standing. In connection with cystinuria the difficulty was still greater, seeing that many cystinurics did not form calculi or develop any conspicuous urinary troubles. Hence, for such anomalies it was practically impossible to construct family trees showing, with any degree of accuracy, the numbers of normal and abnormal members in successive generations. The bearing of the Mendelian theory upon the question of the effects of consanguineous marriages, to which Mr. Punnett had not referred in his paper, appeared to Dr. Garrod to be of extreme interest. The literature dealing with this subject was most unsatisfactory, and most authors had set out to show that consanguineous marriages had or had not evil consequences for the offspring. On the other hand, the explanation that a rare recessive character was most likely to appear in the offspring of the intermarriages of members of a family who produced the recessive gametes seemed to remove the question beyond the zone of prejudice and to explain in a satisfactory manner why so large a proportion of human recessives, such as albinos and alkaptonurics, were the offspring of marriages of first cousins. It also explained the undoubted connection between such marriages and the appearance, in several children of a family, of an anomaly which had not manifested itself in immediately preceding generations.[23]

Garrod refused to become involved in what is now recognized as a sterile debate, and carefully avoided referring to the ancestrian controversy. Major Greenwood vigorously took up the cudgels on the ancestrians' behalf. 'As a pupil of Karl Pearson,' Greenwood thought 'he ought to say something with regard to the Mendelian school':

There being a tendency, apparently, on the part of the Mendelians, to sing a *Te Deum* on the slightest provocation. Not so much in Mr. Punnett's exposition as in the proof of the paper which had been circulated, there was a long list of the conquests achieved by the Mendelian school, and, in face of that, the adherents of that school had no right to complain if criticism were minute in view of its being asserted to be *the* theory instead of *a* theory of heredity ... With regard to night-blindness, in regard to which such a splendid pedigree was exhibited, it was said [by Punnett] to be due to the absence of visual purple in all probability. But later it was [treated by Punnett as] more than a suggestion and was fitted into the scheme. It would be interesting to hear Mr. Punnett's evidence that absence of visual purple was the cause ... Everyone would be glad of proof of the statement.[24]

Greenwood sarcastically suggested that:

The only way of testing it would be to inveigle a subject of night-blindness into a dark room, keep him there two hours, kill him, and then remove the retina and soak it in bile-salts, which, of course, was not done in any of the cases.[25]

Udny Yule, a renowned statistician, weighed in on the ancestrians' side, but provided a mixed and occasionally insightful view:

Many of the cases dealt with by the author referred to the inheritance of abnormalities rather than to disease properly so termed, e.g., such diseases as tuberculosis or insanity. Supposing a definite germinal characteristic decided whether or not a person should have the tubercular diathesis, that did not mean that that man would certainly have tuberculosis, merely that he was liable to have it; he might die of something quite distinct, after living as long as the normal man. And was not that the case in regard to most diseases? The matter was extremely complicated, even if the germinal processes were Mendelian.

[Punnett] said that brachydactyly was dominant. In the course of time one would then expect, in the absence of counteracting factors, to get three brachydactylous persons to one normal, but that was not so. There must be other disturbing factors of equal importance ... [Yule also] doubted whether the theory would at all largely increase the physician's effectiveness in state medicine. On that point Mr. Punnett seemed to be extraordinarily hopeful. Supposing it were found that a certain diathesis was subject to Mendelian principles, did one advance much further, either in treating the individual or in taking general measures? Could the physician do more in the light of such knowledge than he could now? Could he do more than endeavour to keep the individual free from infection and from predisposing conditions? It had been suggested by many writers that the characters were amenable to human control by controlling marriage. That, however, seemed to him [Yule] a chimerical idea, and not in the bounds of the practical at present. Further, in such cases as tuberculosis and insanity, where one could not be certain as to the germinal constitution of the individual, even marriage control would largely break down.[26]

The debate continued interminably, with various speakers extolling both points of view. Sir Shirley Murphy, Vice-President of the Royal Sanitary Institute and a respected epidemiologist, closed the discussion with a statement that he evidently hoped would restore harmony. It did not.

The section ought to thank Mr. Punnett very much for his excellent paper, as the subject matter of it was full of interest. Mendel's law seemed to be established for certain characteristics and within certain limits. How far it might go beyond that was a matter for further enquiry and research. One could not imagine conditions under which Mr. Punnett was going to interfere with the ways of love, but the theory had arrived experimentally at the stage of being of considerable importance to the agriculturalist, and even if it did not go beyond that it would have served its purpose. The interest of that Section in the matter was not as to how far it could be applied, but how far the law was one by which Nature worked.[27]

Punnett wanted the last word. He responded to the criticisms but did not yield on his main thesis. In fact, he could not resist a final dig. Vernon's letter, he said:

raised the old controversy between the Mendelians and the biometricians, and dwelt upon the practical value of the law of ancestral heredity as defined by Pearson and others. But it did not seem … that a law which utterly collapsed before such simple facts as the production of colour from two pure strains of poultry or sweet peas was likely to be of much value to the average medical man or to anybody else.

Mendelian inheritance has now been demonstrated for numbers of most diverse characters in plants and animals. It has also been shown to hold for a few simple cases in man where the evidence has been collected carefully and critically. How far it applies must be a matter of opinion until much more in the way of accurately recorded pedigrees is forthcoming. Facts alone can decide this matter, and if this paper did a little to stimulating the collection of such facts it would have amply repaid whatever pains went to the making of it.[28]

The Mendelians had won the day, but acrimony remained, and the debate would not be settled for many years.

Garrod remained above the fray. The dispute was unseemly and distracting. He was more interested in his 'metabolic sports' and the importance of consanquinity in bringing them about. In this regard, he was not unlike Thomas H. Morgan. A few years before he began to develop the Mendelian chromosome theory for which he was awarded the Nobel Prize, Morgan himself had chided the geneticists, 'pointing out that their game of juggling hypothetical units of inheritance was much too easy and therefore unconvincing'.[29] In the quiet of his fireside, however, Garrod may have contemplated the evolutionary significance of his chemical sports, so many of which were relatively harmless. Did they represent the biochemical variability upon which natural selection might operate? Perhaps that was an even more fundamental biological truth than the outcome of the biological squabble he had just witnessed. He had thought about it before, and would refer to it in his Croonian Lectures in June of 1908.

There was, however, one physician, seven years younger than Garrod, who had become extremely interested in heredity as it applied to medicine. Alfred Milne Gossage had received his undergraduate training in science at Oxford and then entered Westminster Hospital Medical School, graduating in medicine in 1890. Although there are no letters between Bateson and Garrod after 1902, there are a number of letters between Bateson and Gossage that testify to Gossage's fascination with the relevance of genetics to medicine and his reliance on Bateson for advice. Unlike Garrod, who was never much taken with genetics for its own sake, Gossage was interested in the mechanism of heredity, and referred to allelomorphism (the occurrence of two or more variants of a gene at a single locus) as early as 1908.[30] Also unlike Garrod, Gossage was interested in the pedigrees of a variety of malformations. As early as 1907 or 1908, Gossage wrote about heredity in medicine, and when in 1913 Garrod, Thursfield, and Batten edited their textbook of pediatrics, Gossage wrote the introductory chapter on heredity.[31]

During the first decade of the century Bateson and Punnett were preaching the Mendelian gospel at every opportunity, battling with Weldon and the ancestrians whenever the occasion arose. Bateson's scientific association and friendship with Garrod kept him alert to instances where Mendelism could be relevant to man and his diseases; but Bateson recognized the complexity of inheritance, and emphasized that an understanding of heredity would proceed much more quickly through experimental work in plants and animals. With some reluctance he agreed, at Garrod's urging, to address the Neurological Society on Mendelian heredity and its application to man. In this address, delivered in 1906, Bateson spent most of the hour discussing sweet peas, Dorset sheep, and the varieties of primula known as palm-leaved and fern-leaved. He had no doubt that the application of genetics to man 'is rather for the future than the present'. Garrod came in for honorable mention, but although Bateson referred briefly to alkaptonuria at the end of his lecture, he did not mention any of the clinical or biochemical features of the disease. Pedigrees of color-blindness and hemophilia were used to illustrate X-linked inheritance. As an example of an inheritance of a recessive disorder that might be expected to interest neurologists, Bateson selected the example of the Japanese waltzing mouse.

When it came to providing examples of recessive inheritance in man, however, Bateson seemed less sure. He expressed disappointment at the rate of progress: 'the application of Mendelian rules to mankind has not made the progress expected.' He sensed that skin color in man is not inherited in a simple and Mendelian fashion; compared to the inheritance of coat color in the rabbit, the mouse, and the horse, human skin color seemed discouragingly complicated. Surprisingly, Bateson even expressed doubts about the recessive inheritance of albinism, despite its increased frequency among the children of consanguineous marriages. He left his audience with a guardedly optimistic message: heredity is usually not a simple matter; uncertainty abounds in classifying disease as dominant, recessive, or X-linked. Almost as an afterword he concluded:

My object has been to put before you the results of work on lines somewhat remote from your own [primula, sweet peas, and sheep] though as you have seen, presenting points of possible contact. If this discourse may lead to a closer study of the hereditary transmission of disease, I am confident that positive results may be expected.[32]

Although the formation of the Genetical Society was Bateson's idea, Garrod was not among the 26 biologists who gathered at the Linnean Society's rooms on 25 June 1919 to found the new organization; nor is there any evidence that he ever became a member of the society or attended any meetings, even as a guest.[33] Garrod had joined the Physiological Society, probably the premier specialist society of the day, in January 1894 at the early age of twenty-six, and his loyalty to it would endure throughout his professional life. When the Biochemical Society was founded, he decided not to join.

Garrod was, however, a founding member of the Association of Physicians of Great Britain and Ireland. Soon after William Osler arrived in England, in 1905, to become Regius Professor of Medicine at Oxford, he was invited to join a group of senior physicians who were thinking of publishing a new medical journal, one more suited to the needs of academically-minded physicians than the *British Medical Journal*. Osler suggested that the new medical journal should be combined with the creation of an association of physicians, dedicated, like its American counterpart, to 'the advancement of scientific and practical medicine'. In May 1908 Wilmot Herringham invited a number of colleagues sympathetic to the idea to meet at his house in London to discuss Osler's proposal in detail. Garrod was among those invited.

William Hale-White, J. Rose Bradford, Humphrey D. Rolleston, and Robert Hutchison also attended.[34] All five physicians would subsequently be knighted for their services to medicine. They all agreed that the new journal should be an integral part of a new association of physicians, subsequently called the Association of Physicians of Great Britain and Ireland.

The first annual meeting of the Association was held on 23 and 24 May at the Royal Medical and Chirurgical Society in London; 155 members attended. Osler presented a paper on multiple hereditary telangiectasia with recurring hemorrhage. Henry Lewis Jones, Garrod's student companion in Norway, also gave a paper. Although Garrod did not deliver a paper on this occasion, it is highly likely that he discussed Osler's cases.[35]

Osler invited Garrod to assist him in editing the new journal, and in 1907 the *Quarterly Journal of Medicine* came into being. The first paper, on ochronosis, was by Edgar Reid, a surgeon from Swansea Hospital in Wales. Osler discussed the clinical features of the disease, and Garrod discussed the urinary findings in ochronosis. The two contributions neatly highlight the essential differences between Osler, the clinician, and Garrod, the physician-scientist. Osler's account is exclusively descriptive, and the clinical features of the disease are thoroughly discussed. Garrod's section is dry, tightly written, and full of precise chemical information. Garrod described his investigation of the urine of the patient in meticulous detail and provided 13 references (all but two drawn from Continental literature); Osler provided none. All the authors agreed, however, that the case reported was probably a result of the long-term application of carbolic acid to a leg ulcer rather than a late complication of alkaptonuria.[36]

Garrod was a faithful member of the Association of Physicians; he attended the annual meetings regularly, and was elected president in 1923, when the Association met in Oxford. Although he was then sixty-five and had reached the Association's retirement age, his colleagues, citing his remarkable contribution to the organization, elected him to an honorary membership. Garrod's editorial skills were well known, and he was persuaded to remain as one of the editors of the journal for another six years, retiring from the board at his own request in May 1928, having served a 'period of nearly twenty-one years'.[37]

By 1907 the first of many invitations to serve as an external examiner came from his own university, Oxford. Garrod was fifty years old, and firmly established.

The Croonian Lectures:
A Scientific Landmark

Nor can it be supposed that the diversity of chemical structure and process stops at the boundary of the species, and that within that boundary, which has no real finality, rigid uniformity reigns. Such a conception is at variance with any evolutionary conception of the nature and origin of species. The existence of chemical individuality follows of necessity from that of chemical specificity, but we should expect the differences between individuals to be still more subtle and difficult of detection.

— Archibald E. Garrod[1]

THE Croonian Lectures are an honored academic tradition in British science. William Croone, for whom the lectures are named, was born in London on 15 September 1633. In 1659, he was appointed Professor of Rhetoric at Gresham College, London, where he was a colleague of Christopher Wren, who was then Professor of Astronomy. Plans for the creation of the Royal Society were in their final stages, and the founding members met regularly at Gresham College to hammer out the details. In 1662 the Society received its official charter. Croone had been appointed 'Register' in November 1660, and was one of the founding Fellows of the Society. Two years later, by royal mandate, he was created Doctor of Physic of the University of Cambridge. In the same year, he became a member of the Royal College of Physicians, and was later elected a Fellow. Thus it is clear that Croone had two strong academic loyalties: one to the Royal Society, on whose Council he sat, and the other to the Royal College of Physicians. He was an active member of the Royal Society, and read several scholarly papers there, chiefly in the fields of embryology and the physiology of muscle movements.[2,3]

While Croone was undoubtedly a thoughtful scientist, and blood transfusion featured prominently among his fields of interest, some of his communications stretched the credulity of his colleagues, the most noteworthy being his claim that 'a salamander which vomited forth a certain stuff to put out the fire into which it was cast'.[4] In addition to achieving fame as a broadly based scientific scholar, Croone amassed a considerable fortune as a private practitioner in the city of London.

When Croone died in 1684, at the age of fifty-one, he left behind plans for the endowment of two lectureships. One lecture was to be given before the Royal Society, on muscular motion or 'on such subjects as in the opinion of the President for the time being, should be most useful in promoting the objects for which the Royal Society was instituted'.[5] The second lecture, which was to be accompanied by a sermon at the church of St Mary-Le-Bow in the City of London, was to be given to the Royal College of Physicians. There was one small difficulty in the execution of these handsome philanthropic plans, however. When Croone's will was proved, it was found that he had made no provision for the endowment of the lectures. Fortunately, his widow was determined to fulfill Croone's wishes, and began to make the necessary provisions, even after marrying again. In her will of 25 September 1701, she devised to the Royal Society one-fifth of the clear rent from the King's Head Tavern, in or near Old Fish Street, London, at the corner of Lambeth Hill; the remaining four-fifths was to be used to fund the lectures to the Royal College of Physicians.[6] In 1915 the King's Head Tavern was sold to the Corporation of the City of London, and the proceeds were invested. The honorarium for a Croonian lecturer was 10 pounds until 1887 when it was set at 100 guineas; it still is.

The first Croonian Lecture to the Royal Society, read by Alexander Stuart in 1738, was entitled 'Muscular motion'. The first lecture at the Royal College of Physicians was given in 1749 by Thomas Laurence, physician and intimate of Dr Samuel Johnson. The first Croonian Sermon, 'The wisdom and goodness of God proved from the frame and constitution of man', was preached in the same year by Thomas Birch, later Secretary of the Royal Society. The Croonian lecturer was given 'one pound of his paines'. 'The divine for his sermon shall receive one pound,'[7]

The Croonian Lectures to the Royal College of Physicians are second in distinction only to the Harveian Oration, endowed by William Harvey in 1656. William Heberden (1760), Frederick W. Pavy, Garrod's staunch early supporter (1878 and 1894), E. H. Starling (1905), Gowland Hopkins (1927), Henry Dale (1929), Garrod's former student George Graham (1940), and Sir Macfarlane Burnet (1959) are among the distinguished physicians who have delivered Croonian Lectures. The year before Garrod delivered his Croonian Lectures, Sir William Simpson had lectured on plague. Between 1909 and 1990 there have been astonishingly few lectures related to human genetics. In 1971 a lecture entitled 'Nature, nurture, and disease' was given by John F. Brock, and in 1973, Sir John Dacie spoke on the 'The hereditary haemolytic anaemias'. It was not until 1984, 76 years after Garrod's *Inborn errors of metabolism*, that Croonian lecturer David Weatherall, Nuffield Professor of Medicine at Oxford, selected human genetics as his principal theme. The glacial pace of the Royal College of Physicians in recognizing the relevance of genetics to human disease is remarkable. Weatherall, later Sir David, became Regius Professor of Medicine at Oxford in 1992, when he also delivered the Harveian Oration on genetics at the College.

Although Garrod's third and decisive paper on alkaptonuria was published in 1902, it would be another six years before he was invited to deliver the Croonian Lectures, on which so much of his reputation rests today. Garrod made the most of his invitation. Since his observations on alkaptonuria, he had found that albinism, cystinuria, and pentosuria supported his concept of inborn errors of metabolism, and he assembled extensive material on these disorders for the lectures.

Garrod's lectures are now universally recognized as a landmark in medicine, biochemistry, and human genetics. However, there is no record of how they were received at the time by the audience, which consisted not only of Garrod's friends and colleagues, but also of others in the élite of London's medical profession, such as the Council of the Royal College of Physicians, to which he would be elected in January 1909.

When the Croonian lectures were published, in 1909, they were reviewed promptly in the major medical and scientific journals of the day[8] (Fig. 21). All the reviews were short, which is not surprising, since the diseases Garrod discussed were extraordinarily rare, and held little interest for the practicing physician. The unknown author of the review in the *British Medical Journal* provided a somewhat hostile account of the lectures and a muddled understanding of the role of heredity: 'Little is known of their causation [inborn errors of metabolism] but in all there seems to be influence of heredity or of consanguinity.'[9] The reviewer asserted that 'Dr. Garrod maintains the intermediate product theory as the explanation of alkaptonuria, in spite of the experiments which have to some extent militated against this view.' Later, he complains of great uncertainty regarding cystinuria:

In spite of the accepted theory that cystin is produced by the action of diamines, the search for these substances in the urine and faeces is far from being uniformly successful, but Dr. Garrod says that if in any given instance the examination be continued sufficiently long, they may eventually be found ... Dr. Garrod, however, admits that it is impossible at present to form a satisfactory theory of cystinuria.

Referring to pentosuria and its diagnosis, the reviewer wrote:

The failure of the urine to ferment and the failure of fermentation to remove the reducing substance, is a sufficiently trustworthy clinical test, but Dr. Garrod recommends Bial's modification of the orcin test.[10]

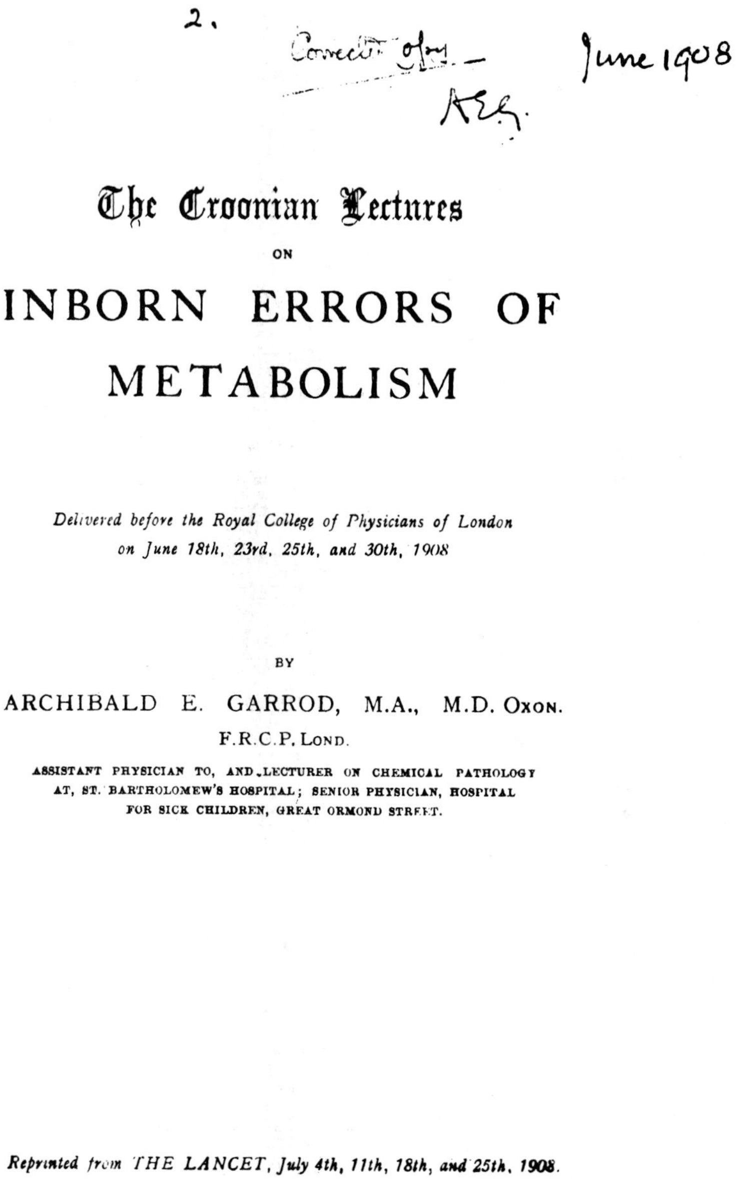

2.

Correct copy — June 1908

AEG.

The Croonian Lectures

ON

INBORN ERRORS OF

METABOLISM

*Delivered before the Royal College of Physicians of London
on June 18th, 23rd, 25th, and 30th, 1908*

BY

ARCHIBALD E. GARROD, M.A., M.D. Oxon.

F.R.C.P. LOND.

ASSISTANT PHYSICIAN TO, AND LECTURER ON CHEMICAL PATHOLOGY
AT, ST. BARTHOLOMEW'S HOSPITAL; SENIOR PHYSICIAN, HOSPITAL
FOR SICK CHILDREN, GREAT ORMOND STREET.

Reprinted from THE LANCET, July 4th, 11th, 18th, and 25th, 1908.

Fig. 21 The Croonian Lectures on *Inborn errors of metabolism;* a corrected copy inscribed by
Garrod in June 1908.

Nowhere does the reviewer perceive that in these lectures Garrod had made a remarkable conceptual leap.

The review in the *Lancet* was considerably longer. The unknown reviewer was more thoughtful, recognizing that Garrod's chemical expertise enabled him to understand the inborn errors of metabolism.

> The whole work is a thoughtful exposition of chemical problems of great complexity manifesting themselves in the sphere of the physician. To the elucidation of these problems the author brings all the knowledge of a thorough master of this branch of organic chemistry, as well as the observant and critical powers of a scientific physician.[11]

The reviewer also grasped that both cystinuria and alkaptonuria may be classified as arrests, rather than perversions, of metabolism. Garrod's concept of biochemical individuality, an essential element of his lectures, was also recognized by the reviewer, who noted that biochemical diversity is just as important as structural diversity, although much more difficult to investigate.

W. D. Halliburton, a friend of Garrod and the author of two classics, *A text-book of chemical physiology and pathology* and *The essentials of chemical physiology*,[12] was allocated one paragraph in *Nature* for his review.[13] He urged everyone interested in metabolism to read the lectures, and he commended Garrod for his originality in drawing physiological conclusions from pathological conditions, rather than taking the more usually adopted reverse approach.

None of the reviewers emphasized Mendelism. Rather, they praised Garrod as a master of organic chemistry. Garrod would not have been displeased with this accolade; his approach to inborn errors of metabolism had indeed been chemical. For the rest of his career he would encourage his colleagues to look at disease from a chemical viewpoint. The Mendelian aspects of disease, apart from neatly explaining the increase with consanguinity, were of less interest to Garrod than the metabolic perturbations associated with human disease.

Garrod's *alma mater* was, perhaps unsurprisingly, alone in heaping praise. The perceptive reviewer in the *St Bartholomew's Hospital Journal* pointed out that Garrod had given new meaning to the notion that 'all flesh is not the same flesh', and had enabled his contemporaries to appreciate better 'some of the differences between the flesh of men, of beasts, of fishes and of birds'. 'We may look forward, in Hopkins' phrase', the reviewer continued, with considerable insight, 'to the time when we shall not speak of a patient possessing a diathesis, but as lacking some intracellular ferment'. The reviewer concluded by remarking that:

Like most pioneer work this book contains hard sayings: it demands and will repay careful study ... All Bart's men will congratulate Dr. Garrod on having written it, and themselves on the fact that such a philosophic and scientific work has emanated from their School.'[14]

Although none of the reviewers, with the possible exception of the reviewer for the *St Bartholomew's Hospital Journal,* appreciated the profound biological significance of Garrod's lectures, the notices were in general favorable. Across the Atlantic, the *Journal of the American Medical Association* had briefly taken notice of the lectures at the time they were delivered, and a review published the year after the lectures had appeared in book form was laudatory, witness the following:

Any physician who wishes a pleasant bit of reading which will give him an insight into the way chemical methods are making progress in medical problems, will find much to interest and please him in this little book, while every progressive physician will value it for its thorough consideration of four obscure metabolic abnormalities.[15]

Although the first edition was undedicated, Garrod dedicated the second edition of the lectures, published in 1923, to his long-time friend Gowland Hopkins. Hopkins undoubtedly admired the meticulously detailed account of the chemistry and intermediary metabolism of *Inborn errors of metabolism.* It is not clear, however, that Hopkins, who had little interest in genetics at this time, differed greatly from his contemporaries in recognizing the biological significance of Garrod's lectures. Hopkins's full appreciation of the genetic and biochemical implications of the lectures was probably awakened when he recruited J. B. S. Haldane, who had been interested in genetics before he joined Hopkins to study the action of enzymes in 1922.

A busy practising physician, Garrod took every opportunity to contribute to the clinical literature. Between 1909 and the outbreak of war, he wrote nine papers, several of them case presentations to the Royal Society of Medicine, which had been formed in 1907 through a merger of the Royal Medical and Chirurgical Society with a number of other medical and surgical societies. Many of Garrod's papers were related to rheumatic diseases and their various manifestations, and consolidated his reputation in the field. These papers, however, have proved more evanescent than his several invited addresses.[16,17] In 1912, the year he became a full physician at Bart's, as well as the year his brother Herbert Baring died, Garrod delivered the Lettsomian Lectures before the Medical Society of London, on glycosuria.[18] Although he was now a Fellow of the Royal Society, he asserted proudly that he was also a 'Lecturer on Chemical Pathology'. He began his lectures by paying tribute to Frederick Pavy, who had recently died. Pavy, one of the few physicians who had appreciated the significance of Garrod's work on alkaptonuria from the very beginning, was an expert on diabetes, and had delivered the Lettsomian Lectures on that subject in 1860. The emphasis throughout Garrod's three lectures was on glycosuria as a metabolic problem. He referred briefly to both alkaptonuria and pentosuria, but not to their inheritance patterns. He also presciently recognized that certain genetic diseases were more common in certain population groups:

There can be no doubt that members of certain races — for example, the Jewish race — are more liable to diabetes than other men ... in some instances, the hereditary tendency is so pronounced as to admit of no question.[19]

Later in the year, when invited to deliver the opening address for the Abernethian Society at Bart's, he returned to a persistent theme, the 'Scientific Spirit in Medicine'.[20] Garrod had read Abraham Flexner's 1910 report to the Carnegie Foundation for the Advancement of Teaching, and he recognized that many of Flexner's criticisms applied to English schools as well as to the medical schools of the United States and Canada. Garrod's lecture to the Society was one of his best. He vigorously exhorted his audience to follow the spirit of science:

Nevertheless, scientific method is not the same as the scientific spirit. The scientific spirit does not rest content with applying that which is already known, but is a restless spirit, ever pressing forward towards the regions of the unknown, and endeavouring to lay under contribution for the special purpose in hand the knowledge acquired in all portions of the wide field of exact science. Lastly, it acts as a check, as well as a stimulus, sifting the value of the evidence, and rejecting that which is worthless, and restraining too eager flights of the imagination and too hasty conclusions.[21]

Garrod strongly encouraged physicians in general practice and consultants alike to embrace the scientific spirit. He firmly believed that the scientific spirit was an attitude of mind rather than a corpus of specialized knowledge. He took time in his lecture to castigate the antivivisectionists, and argued yet again that the laboratory worker and the clinician need to act in concert. He was careful not to minimize the need for clinical excellence in the pursuit of scientific advance: 'personal diagnosis should [not] be replaced by diagnosis by "pink flags" '. He referred to the importance of Hopkins's emerging work on vitamins, which indicated 'that in matters of diet, we have to reckon with obscure factors of which we know hardly anything yet — with items of diet which do not themselves nourish, but which directly or indirectly stimulate nutrition'.[22] The lecture, which did not focus on inborn errors, is remarkable for its insights into the bright future that would attend medical practice if it were grounded in the scientific spirit.

When Garrod became a Consulting Physician at Bart's, he decided to retire from active work at Great Ormond Street. His years there had been thoroughly rewarding, but his work at Bart's, as well as an increasing number of outside commitments, persuaded him, reluctantly, to resign. Laura Garrod, always a strong supporter of the hospital, became a Life Governor, and Garrod's own affection for the hospital would long endure. Garrod was convinced that diseases of children could properly be regarded as a separate branch of medicine. He published in 1913, in association with two young colleagues there, Frederick E. Batten and Hugh Thursfield, the first edition of a notably successful textbook of pediatrics that would remain in print for forty years.[23] It was a fitting conclusion to his years at Great Ormond Street. The last edition, the fifth, was published in 1953, 27 years after Garrod's death.

In this text Garrod stressed that, aside from infectious fever, the principal medical problems in early childhood were 'the pneumococcus, the tubercle bacillus and the organism of rheumatic fever'. Although he refers briefly to congenital malformations, there is no mention of genetics as such. The chapter on heredity was written by Alfred Milne Gossage.[24]

It may seem surprising that Garrod chose not to write the chapter on heredity himself; but he had no wish to become embroiled in the battle between the Mendelians and the ancestrians. Besides, Garrod did not see himself as a geneticist. The genetic aspects of medicine, although interesting to him, were tangential to his metabolic and clinical concerns.

Garrod was quite willing, however, to contribute the chapter 'Disorders of metabolism and diseases of the ductless glands'. In discussing diabetes mellitus, Garrod refers briefly to the role of hereditary influences in the disease. Diabetes was, he thought, an example of a complex metabolic disorder in which control mechanisms had gone distinctly awry. He clearly believed that there must be important hereditary influences at work. In discussing diabetes insipidus, he wrote, 'In some families the malady is strongly hereditary, and may reappear frequently in members of successive generations.' It is likely that Garrod was aware of the papers published by A. Weil in 1884, which indicated that the disease was inherited as an autosomal dominant;[25] but he seemed reluctant to specify the exact mode of inheritance. He was much more at home in discussing alkaptonuria, which he referred to unambiguously as a recessively inherited disease. He readily admitted that cystinuria may appear in successive generations, and stopped short of declaring the disease to be invariably recessively inherited. Garrod's reluctance is fully understandable today, for the disease is now recognized as being genetically heterogeneous.

Throughout the years following the publication of the Croonian Lectures and before the outbreak of the First World War in 1914, Garrod became firmly entrenched in the medical establishment. He had been elected a Councillor of the Royal College of Physicians (1909–11), and was invited to join its select College Club, a dining club that met for evening discussions and fine food once a month. His most prestigious award was his election to the Royal Society on 5 May 1910, at the age of fifty-two. This election had not come easily: Garrod was first nominated in 1903. He received many letters of congratulation, mostly from physiologists and physicians. His old friend Walter M. Fletcher, who was not to be elected until 1915, generously wrote, 'it ought to have happened long ago'. John H. Paddington, echoing Galtonian sentiments, wrote that it emphasized the 'truth of the hereditary capacity'. H. A. Scholberg from the University of Wales quipped, 'Galton and Karl Pearson will be after you for a complete FH!' [family history].[26]

This was not Garrod's first encounter with the Royal Society. He had been in receipt of a small three-year grant from the Society from 1887 to 1890 in support of his work on urobilin, and the deadline for renewal had passed without his having realized it. He wrote apologetically on 21 May 1890 to Sir John Evans, Treasurer of the Society:

This year I accidentally omitted to send in an application for a further grant until it was too late to do so. As I am told that the Council of the Royal Society sometimes makes grants from the donation fund for such purposes, I venture to write to you in the hope that they would possibly be willing to allow me a grant of £15 towards my work.[27]

The Society had forgiven his youthful forgetfulness. Now, twenty years later, they were admitting him to their fellowship. Garrod's father had been elected in 1858, and his brother Alfred Henry, who was first nominated in 1875, had become a Fellow in 1876.

Garrod was nominated by Edward Klein, a distinguished Viennese histopathologist who had been recently appointed to a position at Bart's. Garrod's seconder was William Osler, Regius Professor of Medicine at Oxford. The list of supporting signatures included many fellow physicians with a scientific bent: Lauder Brunton, John Rose Bradford, Frederick W. Pavy, F. E. Beddard, P. H. Pye-Smith, and Arthur Gamgee. Physiology was represented by John Burdon-Sanderson, J. S. Haldane, and W. D. Halliburton. Ray Lankester was also a signatory. Lankester had applied spectroscopy to the study of medicine in 1872, and, like Garrod, was fascinated by pigmented substances, particularly hemoglobin and myoglobin.

From the beginning, Garrod was an active member. He represented the Society on a number of academic occasions, and was a member of the Council twice (1914–15 and 1926–28). He was elected Vice-President in 1926–28, and was elected to the Royal Society Club in 1927. The Club had been founded in 1743, and consisted of some 70 Fellows who met and dined at regular intervals throughout the year. Guests were commonly invited, among them Ivan Pavlov, who spoke, it was reported, in Russian. The guests were nearly always scientists; however, in an unusually festive mood, the Club invited the wine connoisseur André Simon in 1934. Garrod's attendance at Club dinners was irregular, probably because he often had no special wish to journey down from Oxford to what were, after all, largely social occasions. He remained an active member of the Royal Society until 1936, when, because of increasingly failing health, he resigned.

Concern about medical education had been brewing in the United Kingdom for some time, and Garrod also became involved. The appropriate relationship between the University of London and the medical schools had been the subject of discussion since the Medical Act of 1858 became law. Numerous government reports and a recommendation from the Royal College were published, and the reports became increasingly rigorous after the turn of the century. By 1907, the University of London was attracting fewer medical students — one factor leading to the creation in February 1909 of the Royal Commission on University Education, under the chairmanship of R. B. Haldane, subsequently Viscount Haldane. The Haldane Commission's inquiry was massive: it took evidence on 81 days, examined 199 witnesses, asked over 17 000 questions, and produced in all six reports, the last of them in March 1913.[28] The Flexner report on medical schools in the United States and Canada had proclaimed 'the shameless incompetence of the great majority of medical schools', and resulted in the eventual closure of more than fifty of them.[29] Although the Flexner Report was not directed towards medical schools in the United Kingdom, it had a profound and beneficial effect on medical education there.

No physicians served on the Haldane Commission, but the leading physicians from most of the medical schools were invited to provide formal statements of their positions regarding the relationship of the university to the medical school. Many of the witnesses took the position

that, while most of the medical schools provided a sound clinical education, there was woefully little attention being paid to the scientific study of
disease. In this, they concurred with Flexner's findings regarding US and
Canadian medical schools. On 9 November 1911 the views of St
Bartholomew's Hospital Medical School were sought by the Commission.
Sir Wilmot Herringham spoke in general terms of the desirability of
having a professoriate within the school; but he left it to Garrod to
develop his own ideas on the integration of science and clinical medicine.

The commissioners knew that Garrod had studied in Vienna after
graduation from Bart's, and had recently returned from a visit to the
clinic of Friedrich von Müller in Munich. When pressed to discuss the
advantages and disadvantages of German medical education, Garrod
pointed out that in Germany the assistants to a professor were very
experienced, and had often held their positions for many years. Future
professors were recruited from the ablest assistants. No such group of
experienced assistants existed in England. Garrod believed that another
advantage to German medical education was that professors were given
a laboratory in which they and their assistants could pursue scientific
research. As a result, case presentations to students were much more
detailed — by the time the professor was shown the patient, every
chemical test that might have clinical relevance had been performed.
Garrod expressed his view that such a system 'had the enormous
advantage that, apart from the teaching in the clinic, the clinic is a
school of research'.[30]

In 1913 the Seventeenth International Congress of Medicine was
held in London. Garrod's old mentor, Sir Dyce Duckworth, was
honorary president, Sir William Osler was president, and Garrod himself
was one of the two principal organizing secretaries, and delivered a
major address at the Congress, on the clinical application of pathological
chemistry.[31] He discussed the urinary findings in pentosuria, cystinuria,
and alkaptonuria, but did not indicate that these anomalies, as he called
them, were inherited. He mentioned some of his more recent enthusi-
asms, such as hematoporphyrinuria, which gave him the opportunity to
discuss the clinical usefulness of the spectroscope. In closing, Garrod
told his audience that the motto of the old Pathological Society, 'Nec
silet mors', ['Death is not silent'] was out of date. 'It is upon the cordial
cooperation of the ward and the laboratory that the pathology of the
future must be built up.'[32] Gowland Hopkins, in a major address
entitled 'The dynamic side of biochemistry', emphasized the importance
of studying the biochemistry of the cell, and referred specifically to
Garrod's contribution.[33] William Bateson spoke on heredity.

In addition to the scientific program, there were a number of social occasions associated with the Congress, and Garrod had taken Laura and their daughter Dorothy, who was twenty-one and reading history at Newnham College, Cambridge, with him. Friedrich von Müller, a friend of Garrod's for many years, was staying at Brown's Hotel, within easy walking distance of the Garrod home at 10 Chandos Street. Despite the international nature of the Congress and the predominant optimism that the difficulties between Germany and her neighbors were being slowly resolved, the clouds of war were gathering over Europe. It would be more than five years before Garrod resumed his normal medical and social friendships with his colleagues on the Continent.

8

Malta: The War Years

WITH the declaration of war on 4 August 1914 Garrod, like so many of his colleagues at St Bartholomew's, put aside his private consultant practice, as well as his academic concerns, and joined the armed forces. For the first fifteen months of the war, he served, with the rank of major, as a general medical consultant at the First London General Hospital in Camberwell. In November 1915 he left England and sailed for Malta to join the Mediterranean forces as a temporary colonel in the Army Medical Service[1,2] (Fig. 22). Earlier, in July 1914, Garrod had

Fig. 22 Colonel Archibald Edward Garrod (second from right) and fellow medical officers in Malta.

delivered the major medical address, 'Medicine from the chemical standpoint' at the eighty-second annual meeting of the British Medical Association, in Aberdeen, Scotland,[3] and had received an honorary doctorate of laws from the University of Aberdeen.

Garrod was pleased to be posted to Malta to serve as a Consultant Physician in the military hospital at Gozo. The war had already cost the Garrods their son Thomas, who was killed in action earlier in 1915; and Alfred was serving in France. Garrod felt too removed from the war, living at home in the comparative luxury of Chandos Street. Moreover, the patients he saw at the London General Hospital did not differ greatly from those he would have seen at Bart's. Since early childhood Garrod had been fascinated by geography and the ecology of small islands; at the age of sixteen he had accumulated sufficient information to give a talk on islands to his fellow members of the Natural History Society at Marlborough.

Garrod's posting to Malta not only enabled him to serve his country overseas, but also provided him the rare opportunity of living on an island rich in history and archaeological interest. Variously held by Phoenicians, Greeks, Carthaginians, Romans, Arabs, and Spaniards, the island was turned over in 1530 to the Knights of St John of Jerusalem, also known as the Knights of Malta, who made it a stronghold of Christianity, and successfully checked the advance of Islamic power in southern and western Europe.[4] The Knights were expelled by Napoleon in 1798, and the Maltese people requested the protection of the British Crown in 1802. In 1814 General Sir Thomas Maitland proclaimed the intention of King George III of England to recognize the people of Malta and the neighboring island of Gozo as subjects of the British Crown.

When war broke out Malta was a peacetime garrison on a small island of just under 250 000 inhabitants, with four military hospitals and a small medical establishment. Unfortunately, this tranquil atmosphere was not to last. On 25 April 1915 a British expeditionary force landed at Gallipoli; by the end of May 4000 sick and wounded had been sent to Malta, and the number of hospital beds had increased to 7000. When Garrod reached Malta in November the sick and wounded were arriving at the rate of 2000 per week.

As soon as Garrod arrived, he took over the medical duties of Colonel G. L. Gulland. Gulland, a consulting physician to the Royal Infirmary in Edinburgh, had joined up as soon as war was declared, but was now returning home to continue his university duties.

Although Laura Garrod accompanied her husband to Malta, she stayed for only a short while, as did most of the wives of officers. During the war she returned for two brief visits to the island, in 1916 and in 1918. Garrod returned to England in the summer of 1916 to receive the Companionship of the Order of St Michael and St George (CMG), a decoration instituted by George III in 1818 to commemorate the transfer of Malta to British rule. Apart from these brief periods, Garrod was separated from his family from December 1915 until January 1919. When Laura returned to England in 1915 Garrod stayed with James Purves-Stewart, whom he had known in London, and his wife.[5] An eccentric neurologist and a near-contemporary of Garrod, Purves-Stewart was a man of strikingly different personality. He had served with distinction in the South African campaign of 1900–1, and was aggressively proud of his military service. His ward rounds was notable for histrionic gesticulation and unstoppable volubility, in stark contrast to those of Garrod, whose competence was often obscured by his gentle, self-effacing manner. In about June 1916 Purves-Stewart, who had been created Companion of the Order of the Bath (CB), and who two years later would be knighted, Knight Commander, Order of St Michael and St George, (KCMG), was posted to Salonika and replaced by Colonel Howard H. Tooth.

Tooth was an old friend of Garrod's from Bart's. Like Purves-Stewart he was a neurologist and had served in the Boer War; but his personality was much more in tune with Garrod's. A cheerful and popular man, Tooth shared many of Garrod's cultural interests, including a great fondness for music, and they got on well together. In June 1916 Garrod moved into a house which he shared with Tooth and Charles Ballance in the Piazza Miratore in Floriana, a suburb of Valletta, the capital of Malta. Ballance, a well-known and erudite surgeon with a fondness for quoting Shakespeare, had worked with Purves-Stewart on the surgical treatment of neurological injury. When Ballance returned to England Laura Garrod and Howard Tooth's wife lived in Floriana for a few months. Later, Garrod and Tooth left Floriana, and made their home in the elegant Osborne Hotel in Valletta.

Garrod settled in surprisingly quickly. Malta fulfilled all his expectations of what an island, and an island people, should be. In his farewell lecture on 21 January 1919, in the Aula Magna of Malta University, founded in 1769 by the Knights of Malta, and from which diplomas in medicine were granted a few years later, Garrod expressed his love for islands and for Malta in a style that verges on the poetic:

Every child loves an island and peoples it with creatures of his fancy. It may be that this fascination is a legacy from a very remote past, for the simple workings of the minds of children reflect, to some extent, the thoughts of primitive man. The attraction is not wholly lost as the child grows older; and even in later life most of us are still conscious of the island charm.

He who first comes to Malta in the early morning can hardly escape it, as he sees the shadowy outlines of the island emerge from the horizon mists, the unfamiliar features and colouring of the nearer landscape, the harbour girt by the bastions and half-oriental buildings of Valletta and her sister towns, the gaily-coloured *dghaisas* which flock around the ship. Perhaps the charm has more appeal for us islanders than for mainland folk, but I am assured that they, too, are not immune.[6]

Garrod's previous medical experience, although broad, had not provided him with any special knowledge of diseases of the tropics, and he was initially ill-prepared to diagnose and treat many of the conditions prevalent among the troops serving in the Mediterranean theater. In addition, the sheer number of patients needing attention was daunting: in a single week in December over 6000 sick and wounded were landed in Malta for medical care. By February 1916, however, owing to the evacuation of the forces from Gallipoli, the numbers decreased sharply, and for about four months Garrod had more time to himself. This state of affairs did not last.

In July 1916 an outbreak of malaria among the troops in Macedonia (eastern Turkey) greatly increased the medical responsibilities of Garrod and his colleagues. Many more beds were needed, and the authorities responded quickly. Within three months an additional 25 000 beds were set up in schools and barracks to accommodate the influx of patients.[7] Garrod, always a hard worker, took the increased workload in his stride, unceasingly traversing the island giving sensible and reassuring advice to the inexperienced young medical officers.

Among the many obscure clinical problems that faced Garrod, the various forms of dysentery that afflicted the troops were particularly perplexing. The diagnostic challenge of sorting out these puzzling illnesses appealed to him, however, and he soon became clinically proficient in handling a variety of complex, often ill-understood diarrheal diseases. Although most of these diseases were of bacterial origin, some were caused by the parasite *Entamoeba histolytica*, which causes the clinical syndrome known as amebiasis. The parasite was commonly found in the contaminated water and food that were too often a feature of wartime conditions in the Mediterranean. In heavy infestations, the parasites migrate via the bloodstream to the liver, where they can cause life-threatening suppurative liver abscesses — the amebic abscesses. An amebic liver abscess could be extremely difficult to detect, and Garrod impressed on his medical officers the diagnostic usefulness of careful percussion over the liver if an abscess was suspected. Because malaria was rife among the troops, the differentiation between splenic enlargement caused by malaria and that due to leishmaniasis, more commonly known as kala-azar, was exactly the kind of intellectual challenge that Garrod relished.

In February 1916 he was joined by George Graham, a promising young doctor who had been Garrod's house physician at Bart's. Graham stayed in Malta with Garrod until April 1917, and remained a devoted disciple and friend throughout his life. One of his responsibilities was to accompany Garrod on his lengthy ward rounds. Although Graham admired Garrod's keen diagnostic acumen, he expressed the view that the treatment of disease did not hold the same fascination for Garrod as arriving at the correct diagnosis. It was not that Garrod was unsympathetic or uncaring, but as a physician steeped in the scientific tradition he was reluctant to adopt poorly substantiated therapeutic claims as readily as some of his more clinically oriented colleagues. Garrod's reputation as a good medical opinion soon spread to the civilian population, and he was frequently consulted by wealthy and prominent Maltese families. They appreciated his thoughtful medical advice and gentle nature, even when unaccompanied by the polypharmacy so often adopted by his contemporaries and expected by many of his patients.

Graham was later to remark that, for Garrod:

visiting a different hospital each day of the week, and very occasionally on Sunday, made it very difficult [for him] to remember all the cases and I was able to help him here and also to remind him about treatment. For he [Garrod] was rather like the great Bartholomew's physician, Dr Samuel Gee, much more interested in the diagnosis of disease than its treatment, and it was for this latter that the Medical Officers often wanted his advice.[8]

Graham was impressed by the rapidity with which Garrod, although hampered by the absence of reference books in Malta, learned about kala-azar, Malta fever, malaria, and the various dysenteries.

The number of patients arriving in Malta began to decrease as the disastrous campaign in Mesopotamia drew to a close; but eruption of the war in Salonika led to an additional influx. G. R. Bruce estimated that 2600 officers and 64 500 men of other ranks were evacuated to Malta for treatment as a result of the murderous fighting in Salonika.[9] Later, as the intensity of submarine activity in the Mediterranean increased, it was considered unsafe to move the troops from Salonika to Malta, and the number of new casualties decreased dramatically. Although there were 350 army medical officers on the island, there were only three consultant physicians, one of whom was Garrod. In an effort to share their unaccustomed experiences, the medical officers gathered fortnightly for a conference to discuss the most perplexing patients under their care. Garrod faithfully attended these meetings, and was a source of reassurance and encouragement to the younger physicians. At the fortnightly meeting on 5 January 1917 Garrod discussed the condition known as 'soldier's heart'.[10]

'Soldier's heart' was a diagnostic label frequently given to a cluster of symptoms suggesting a disturbance in circulatory regulation. Lassitude, weakness, and a feeling of faintness were characteristic, and the syndrome was commonly encountered among soldiers suffering from prolonged physical and mental strain. Garrod had observed that many of the men with so-called 'soldier's heart', as well as those with malaria, were not only short of breath, but also often developed a slight enlargement of the right side of the heart, as judged by percussion. Garrod maintained that careful palpation could often reveal slight enlargement of the heart. He believed that by dipping his fingers along the course of the ribs and intercostal spaces he could detect an increase in resistance in patients with enlargement of the right auricle. This uncertain claim was the more surprising because it was expressed with an uncharacteristic forcefulness that bordered on the bellicose. In a paper for the *Lancet*, he wrote: 'I am well aware that some physicians, whose opinions are entitled to all respect, deny the possibility of determining the right border of the cardiac area by percussion, let alone palpation, but this is wholly at variance with my experience.'[11]

Garrod's insistence on the usefulness of dubious physical signs seems quite out of character with his customary analytical and scientific approach to diagnostic problems. But in Malta his clinical skills had been well honed and well tried. As Consultant Physician, he felt personally responsible for the clinical care of the soldiers on the island, and probably did not particularly welcome the views of disputatious young medical officers.

In the *Lancet* paper, he insisted that right-sided cardiac enlargement with accompanying symptoms of shortness of breath, whether due to malaria or soldier's heart, should always be treated by strict bed rest; he believed that soldier's heart was not a distinct clinical entity, but encompassed a variety of symptoms whose treatment was bed rest. This general point of view is now widely accepted, and the term remains only as a historic reminder of a condition that was seen among the troops in the First World War and for which no unitary cause could be found. Garrod was convinced that his patients would not only benefit greatly from rest, but would be made distinctly worse by exercise — a treatment often misguidedly advocated for patients with the condition.

In June 1916 an old friend Arthur Hurst, Consultant Physician to Guy's Hospital, stopped in Malta on his way home from Salonika, where he had been stationed, to visit Garrod. He stayed for two weeks, accompanied Garrod on his daily rounds, and had an opportunity to observe Garrod closely as he examined his patients. Hurst was later to comment in his autobiography that the view expressed by Francis Fraser in Garrod's obituary notice — that Garrod was a great biochemist but no clinician — was quite untrue. 'What I saw of him in Malta showed what a fine general physician he was, although at home his chief interests were in the investigation of diseases in which the chemistry of the body fluids was disordered, such as gout and diabetes.'[12] It is quite likely that while Hurst was in Malta he and Garrod discussed the possible role of genetic factors in the causation of disease, and the general question of chemical individuality. However, to judge from Hurst's own lecture on 'constitutionality' in 1927, an appreciation of the role of genetics, except for segregating Mendelian traits, eluded him, as it did many other physicians of the day.[13] Hurst continued to discuss constitutionality in unsatisfactory and vague terms, and he seems not to have recognized fully that 'constitutional factors' were 'genetic factors', not part of some mysterious, indefinable diathesis. Although some of the diseases Garrod discussed, such as diabetes and gout, were close to Hurst's own interests, and their discussions were mutually advantageous, Hurst was unable to grasp the biological meaning of constitutionality and its relation to overt disease.

Among the physicians posted to Malta during Garrod's tour of duty was a brilliant young medical officer Charles Putnam Symonds, who became one of the most distinguished clinical and academic neurologists of his day. Symonds's father, Sir Charters Symonds, was a distinguished surgeon; and he, too, stayed with Garrod for the brief time he visited the island. Although Charles Symonds was more than thirty years younger than Garrod, they soon established a close friendship, and Garrod invited the young man to stay with him. When free from their demanding duties, they took long walks together around the island. Their conversation ranged widely, and Symonds learned from Garrod much about the prehistory of Malta.

One of Garrod's lesser-known interests, dating from his school-days, was archaeology. In Malta, he had a unique opportunity to learn about the subject firsthand. He developed a friendship with a Maltese physician Professor (later Sir) Themistocles Zammit, who had worked with Sir David Bruce on the elucidation of Malta fever and who had helped make the discovery that goat's milk was the main source of the infection. (Indeed, in 1964 Malta issued two stamps honoring their joint achievements.) Zammit, in addition to being a skilled bacteriologist and physician, was a keen and talented archaeologist who held the position of curator of the Malta Archaeological Department. Whenever Garrod could escape from his busy schedule he would accompany Zammit to his digs, where he participated in the excavation of the recently discovered Tarxien Neolithic temples, which indicated the presence of a Bronze Age people around 2000 BC (Fig. 23).

Under Zammit's friendly tutelage Garrod enthusiastically learned all he could about Malta's prehistory. He eventually became so knowledgeable that Zammit asked him to read an early manuscript of his book *Malta: the islands and their history.*[14] Although it seems unlikely that Garrod was able to add materially to the technical aspects of the treatise, his love of the English language and his insistence on clarity and felicity of expression undoubtedly improved its readability.

Another colleague of Garrod's in Malta from 1915 to 1917, and also a friend of Zammit's, was Charles Singer, a physician who would become an eminent historian of medicine. Singer also accompanied Zammit on his archaeological excavation at the Tarxien Neolithic temples, and in 1924 co-authored a paper with him.[15]

Garrod had a strong hunch that archaeological research would advance more rapidly if archaeologists used the tools and methodology

Fig. 23 Garrod engaging in an archaeological exploration in Malta, one of his favorite diversions while stationed there.

of science. The physical sciences had shed so much light on the understanding of many medical problems; might not archaeologists do well to apply them to their scholarly activity? In Garrod's lecture *The University of Utopia*, delivered at the University of Malta on 3 November 1917, he described the life and work of an archaeologist with such enthusiasm that it would have been easy to imagine archaeology was his profession rather than an avocation.[16]

It has often been asserted that Garrod was by nature shy, with little or no sense of humor. This is only partly true. There is good evidence from those who knew him that his lectures were often laced with irony and a quiet wit. In *The University of Utopia*, he allowed himself a certain wry humor: 'Even the sensibility of idle hands upon the walls helps to reveal the daily life of remote peoples, and it is a comfort to think that those who so deface historic spots nowadays, may be providing valuable graffiti for archaeologists of the future.'[17]

Although Garrod did not sparkle as a bedside teacher, his formal lectures, even on scientifically complex subjects, were carefully crafted, scholarly, and informative. He wore his deep learning lightly, and had an elegant style that at times bordered on the poetic and could hold an audience rapt. While in Malta, he was often invited to give informal lectures to lay audiences. Four of them were published, and still stand as models of medical and literary exposition.

The University of Utopia was the first of these lectures to be published. In it, Garrod emphasized the importance of basic science to clinical practice, and focused on what he saw as the crucial difference between a technical school and a university. Few have expressed it better:

But to convert our universities into higher technical schools, to the exclusion of the pursuit of knowledge for its own sake, would be a disastrous step, nothing less than the killing of the goose which lays the golden eggs. The wonderful advances of applied science, which have transformed the world, have sprung from the patient investigations of men who have searched out the secrets of nature for sheer love of knowledge, and with no eye to its practical application. When Galvani investigated the contractions of the frog's legs prepared for his dinner, as the breeze blew them against the bars of his balcony, and so laid the foundations of our knowledge of current electricity, he can have had no vision of the future triumphs of electrical engineering which should follow from his discoveries.

The main functions of the University are four in number. The more important are the instructions and research, the less important, examination and the bestowal of degrees.

Just as pupils originally gathered to the informal teaching of eminent men, renowned for their erudition, so nowadays the university system calls for professors who possess such knowledge as is only to be acquired at first hand. They should be investigators who are still pursuing their investigations, whether the field of their studies be the lofty concepts of theology and philosophy, the minutiae of grammatical form, the foundations of history, or the study of the phenomena of nature by observation and experiment.

None of us can hope to be more than partly educated, and, if one has to choose between the two, the ignorance of natural science at present prevailing is far more deplorable than would be the disappearance of Greek from the curriculum of other than classical students.

Nevertheless an education which is wholly scientific is just as lop-sided as one from which science is excluded; for there are great fields of thought which are not controlled by scientific reasoning, but, shorn of which, man would lose many of his highest attributes.[18]

Pungently, he continued to give his opinion of what science was all about:

Science is not, as so many seem to think, something apart, which has to do with telescopes, retorts, and test tubes, and especially with nasty smells, but it is a way of searching out by observation, trial, and classification; whether the phenomenon investigated be the outcome of human activities, or of the more direct workings of nature's laws. Its methods admit of nothing untidy or slipshod, its keynote is accuracy and its goal is truth.[19]

Garrod delivered a second lecture, entitled *Other worlds than ours*, to the Church of England's Men's Society at the Valletta gymnasium on 11 February 1918, but unfortunately there is no record of its content. The lecture was illustrated by slides, and Frank Vella, who more than anyone else has studied Garrod's years in Malta, believes that it was an archaeological rather than a scientific talk, and that slides were provided by Zammit.[20] It would certainly be consistent with Garrod's orderly mind that he would want to summarize what he had learned about the archaeology of Malta before leaving the island.

The day after the armistice was signed, Garrod gave his last lecture in Malta, also organized by the Church of England's Men's Society, and entitled *Life on other worlds*. Garrod was far from being a cosmological mystic and it is frustrating not to know how he addressed the subject of extraterrestrial life or how his views would have compared with J. B. S. Haldane's in the essay *Possible worlds*, published twenty years later.[21]

Although Garrod's days in Malta were medically and intellectually rewarding, and he made many new friends, they were clouded from the beginning by an irretrievable loss, the death of his son Thomas Martin in 1915 of wounds received in action in France, near Richebourg St Vast. Thomas was only twenty years old. He had joined the army at the beginning of the war, and had been commissioned as a lieutenant in the 3rd Battalion, Loyal North Lancashire Regiment (Fig. 24).

When Garrod arrived in Malta later in 1915, his eldest son, Alfred Noël, was twenty-seven years old (Fig. 25). Alfred had attended Marlborough College and Emmanuel College, Cambridge, and had obtained a pass in the natural science tripos in 1909. He entered Bart's as a medical student, and shortly after graduation in 1914 enlisted in the Royal Army Medical Corps as a medical officer. He was commissioned in July 1915, and was sent to France in November with the 100th Field Ambulance. On 25 January 1916 he was killed at Givenchy by a shell that hit the ambulance unit to which he was attached. He had been in

Fig. 24 Thomas Martin Garrod, Garrod's second son, died of wounds received in action in France in May 1915.

Fig. 25 Alfred Noël Garrod, Garrod's oldest son, was killed at Givenchy in 1916 by a shell that hit the ambulance unit to which he was attached.

France two months, and was barely twenty-eight years old. Alfred had written a short article about life as a regimental medical officer for the hospital journal. His last sentence has a special poignancy: 'If you wish to do medicine or surgery don't be a regimental M.O.; if you want to see life, do.'[22]

Both sons were laid to rest in the town cemetery at Béthune in the Pas de Calais. Following a request from their mother, a memorial tablet was placed on the north aisle of the parish church of St Andrew, Melton, close to Wilford Lodge, the ancestral home of the Garrods (Fig. 26), on 14 September 1918.

From every account, Garrod's enthusiasm was for ever quenched, and he became a reserved and private person. The profound effect of the war on Garrod has been vividly captured by Purves-Stewart:

He [Gulland] was succeeded by Archibald Garrod of Barts, a delightful colleague and a man of wide culture. I remembered him from the old days, when I was a house-physician at Queen Square and how Garrod used to come in, from the adjoining children's hospital in Great Ormond Street, to investigate obscure bio-chemical problems. In those days he was a strikingly handsome man with leonine head, jet-black hair and deep, glowing eyes, with a charming, kindly manner. When he arrived in Malta his appearance had dramatically changed; he was now white-haired, with the same deep-set eyes, and a pathetic sad expression. He had already lost one son in the War. During his term of service with us in Malta he received the news that another son had been killed. To this fresh bereavement he bore up with great fortitude.[23]

Fig. 26 Wilford Lodge, Melton, Suffolk, ancestral home of the Garrods.

Garrod's only daughter, Dorothy Annie Elizabeth, was twenty-two years old when war was declared (Fig. 27). She attended Newnham College, Cambridge between 1913 and 1916, and read for the history tripos, achieving only a class II.2 degree, a surprisingly low pass considering her later illustrious academic achievements. Her relatively poor showing at Cambridge was ascribed by her close friend Gertrude Caton-Thompson to her anguish at the recent deaths of two of her brothers. As soon as possible after graduation, and after a short period in the Ministry of Munitions, she enrolled in the Catholic Women's League, and served in France and Germany. Although Dorothy's conversion to Roman Catholicism initially disturbed her father, her newly found religion proved a great and lasting comfort. Her experience with the League in France enabled her to be more linguistically comfortable with French archaeological colleagues with whom she worked, and also led to close friendships with the archaeologist Abbé Henri Breuil and paleontologist Pierre Teilhard de Chardin.

Fig. 27 Dorothy Annie Elizabeth Garrod who, by the age of 26, had lost her three brothers and was the sole surviving child of Archibald and Laura Garrod.

At Easter 1916, in an effort to prevent Dorothy from dwelling too much on the deaths of her two brothers, her parents invited her to Malta. There she was introduced to Zammit and a view of prehistory that stimulated her childhood interest in antiquities. Zammit asked Dorothy to review the second edition of his book. As a young girl, Dorothy, after the usual nursery upbringing, had been tutored by Isobel Fry, later Mrs John Masefield, wife of the Poet Laureate and one-time casualty officer at Bart's. As part of her tutoring, and to relieve the monotony of book work, Isobel had taken Dorothy to St Alban's Cathedral, and together they had visited other Gothic churches. When Dorothy became a distinguished archaeologist she would often refer to her first visit to the eighth-century cathedral of St Alban, and the influence it had in shaping her future career.

On 16 December 1916 the University of Malta conferred degrees of MD, *honoris causa*, on Garrod and his colleagues Charles Ballance, Walter Thorburn, and Howard Tooth in recognition of 'their professional eminence and the important work they are rendering in connection with the war'. The degree ceremony was a gala affair presided over by Field Marshal Lord Methuen, then Visitor to the University. A generous press heaped praise on the recipients of the honorary degrees:

The beautiful hall [at the University] from the walls of which look down the grave and revered countenances of several generations of Maltese worthies, was filled throughout ... In front of the dais sat the four gentlemen to be honoured, square soldierly figures in khaki, and splendid types of British manhood as well as the noble profession which they adorn.[24]

Garrod, although undoubtedly pleased with this warm recognition of his contributions, must have been somewhat embarrassed by the excessive adulation. He was not given to ostentation, and as he sat in his place of honor he must have been thinking of his two sons and their tragically brief lives.

On 11 November 1918 the war that had cost nine million lives and had claimed two of Garrod's three sons finally came to an end. Both Garrod and his son Thomas Martin had been mentioned in dispatches during the war, and in 1916 Garrod had been appointed Companion, Order of St Michael and St George (CMG) by the King. In 1918 he became a Knight of the same Order.

Garrod was ready to leave the island that had been his home for the last three years, and to which he had become so attached. The Maltese people appreciated him not only because of the superb medical work that he so tirelessly supervised, but also because he had taken the trouble to learn so much about their island. From Zammit and his archaeological friends Garrod had acquired a deep knowledge of the island's prehistory, and from his many walks around the island, often accompanied by Graham and Symonds, he had learned about Malta's abundant flora. He was probably introduced to the curious Maltese fungus, formerly so valued for its medicinal purposes that, under the rules of the Knights of Malta, a guard was set up to preserve it.[25]

On 2 January 1919 a farewell dinner was held to honor Garrod and Lord Methuen,[26] who had received an honorary degree earlier in the day. The toast to Garrod bespoke the feeling of the Maltese people:

Garrod is a household word not only among gentlemen belonging to the medical profession but even laymen like myself. Colonel Garrod is further entitled to the gratitude of us, Maltese, for his kindness and generosity towards the civil population who have never appealed to him in vain for his medical assistance, and who have benefitted by his kind and disinterested attention.[27]

The final toast of the evening struck an even more personal note: 'Gentlemen, it is to the eminent physician, to the eminent citizen, to him who has so cruelly suffered by the war and who is now one of our

Fig. 28 Basil Rahere Garrod, Garrod's youngest son, died of influenzal pneumonia in Germany just three months after the Armistice was signed.

Fig. 29 Memorial crosses to the Garrod sons and other victims of the First World War in Melton Old Church. By kind permission of Mark Ward – Elm Studios.

real friends, that I ask you to lift up your glasses.' In replying to the toast Garrod 'confessed that the loss of his two sons was most painful, but even in such pain, there was a certain amount of pride which time could not efface, the pride of the sacrifice and the pride that those most dear to him had done their duty and had fallen for a just cause'.[28]

On the eve of Garrod's departure for England, the *Daily Malta Chronicle* saluted him with the following valedictory remarks:

We shall certainly lose a most prominent, familiar and sympathetic personality, as not only will a gifted scientist and distinguished physician of world-wide reputation disappear from our midst, but also one of the best types of the perfect English gentleman. Colonel Garrod, to give him the name by which he is best known amongst us, soon commanded general respect, esteem and admiration. He may indeed say with Caesar: *Veni, vidi, vici,* for he came, travelled over Malta a great deal more than the average Maltese could claim to have done and easily won the hearts of us all. It was predicted that he will not easily forget the little Island which was so dear to him, its University which he honoured and its inhabitants generally who had the privilege and fortune of largely benefitting by his vast knowledge and experience, during his four year stay among them.[29]

Garrod left Malta, the island that had embraced him as one of its own and where he had had much joy as well as sorrow, resigned his temporary commission in the Army Medical Service, and returned to his old hospital.

His relief that the anguish of war was over and that he could resume his work at Bart's was short-lived. The War Office was soon to inform him that his last surviving son, Basil Rahere (Fig. 28), had died. Basil Rahere had attended the Royal Army Military Academy at Sandhurst and enlisted with the Loyal North Lancashire Regiment in 1916. In January 1917 he had joined the 1st Battalion, which was already in France. After

Fig. 30 Colonel Garrod, his uniform bearing a black armband in mourning for his sons.

serving in France for only a short while, he contracted trench fever. On his recovery, he was attached to the Royal Air Force as an observer, and returned to France to join the 149th Squadron, with which he served until the Armistice. After the Armistice, he went to Germany with the Army of Occupation. Three months after the Armistice was signed, and while waiting in Germany to be demobilized, Basil fell victim to the virulent influenza epidemic that killed more than twenty million people worldwide.[30] He died in Cologne on 4 February 1919 of influenzal pneumonia on the eve of his demobilization. He was twenty-one.

With their cherished hopes for their last surviving son shattered, Archibald and Laura arranged for an additional memorial plaque to be placed on the wall of the Church at Melton. There beside his deceased brothers, Alfred Nöel and Thomas Martin, Basil Rahere was laid to rest (Fig. 29).

Garrod never recovered from the loss of his three sons, with all their youth and promise (Fig. 30). He immersed himself in his work, and talked little to his friends and colleagues about his personal tragedy. His sense of loss was deep and irreconcilable. For Archibald and Laura Garrod, nothing would be the same again.[31]

The Call to Oxford

GARROD had been greatly missed by his colleagues at Bart's who had remained behind to ensure that the hospital continued to serve the people of London. The *St Bartholomew's Hospital Journal* in April 1919 took special note of his return from overseas: 'Sir Archibald has been doing most valuable work in Malta, and we extend a warm welcome to him after his long absence'. Garrod, though visibly depressed, was pleased to be back at his old hospital, renewing acquaintances, giving lectures, and teaching medical students. Medical education and medical research had been greatly stimulated by the work of the Haldane Commission before the war. Garrod had argued before the Commission, as well as in his general lectures, that there was a vital need for closer collaboration between physicians and scientists, and that an opportunity must be provided for clinicians to pursue their own research.[1] A special medical unit within the hospital, led by a full-time director and with clinical research as its primary goal, would meet these needs.[2]

An invitation to participate in a symposium to honor William Osler, on the occasion of his seventieth birthday, 12 July 1919, gave Garrod a suitable opportunity to express his views on the relationship between the laboratory and the ward. He assured his audience that there were many opportunities for the practising clinician to advance scientific knowledge. It would be a tragic mistake, he stressed, to leave the care of patients entirely to clinicians uninterested in advancing knowledge, while scientific advance was left to scientists remote from clinical responsibility. He pointed out that:

... the entire groundwork of our modern knowledge of internal secretions was laid by practising medical men, amongst whom were our own countrymen Addison, Graves, Gull, and Horsley, who gained an insight into the functions of the ductless glands by studying the experiments which nature is always carrying out before the eyes of those who have eyes to see them ...

... there is still much work ahead for clinical physicians and surgeons, in the advancement of knowledge as well as in the treatment of the sick, and especially for concerted work in which laboratory and ward workers co-operate as colleagues, and without any claim to a monopoly of the scientific spirit and method on either side ...

If it were possible to concentrate the whole of the scientific work of medicine in the laboratories, and if clinical studies came to be regarded as unattractive to men of scientific instincts, the results would be deplorable for medical practitioners and patients alike. Even now there are signs of diminished zeal on the part of student to become adept at the purely bedside methods of examination. Unless the whole field of medicine be permeated by the scientific spirit we can look for little progress, and shall return to the conditions in the Middle Ages, in which practitioners merely accepted and applied what was to be learnt by study of the writings of Hippocrates and Galen.[3]

Seventy years later, these words remain among the most eloquent and persuasive statements of the need to create opportunities for clinicians to contribute to basic knowledge through their research. The need for clinical research could be met, Garrod believed, by a special allocation of hospital beds, where particular attention could be paid to the pursuit of new knowledge. He had already expressed clearly and unambiguously his conviction that this goal could be achieved most easily by the creation of professorial units, and he believed that it would not be difficult to introduce such a system into the existing hospital structure. There need be no reduction in the emphasis on clinical bedside teaching, nor was there any need, in his opinion, to adopt the tradition of a formal daily clinical lecture.[4] Departing from Osler and the Continental tradition, Garrod firmly believed that the professor in charge of the unit should be employed full-time, and should not engage in private practice:

The unit plan is quite compatible with our system of small cliniques under the physicians and surgeons to the hospitals, but the introduction of professorial cliniques *of the continental type* would involve more sweeping changes. Such a clinique is a much larger unit, in which the staff consists of a professor with a large number of beds under his care, and a group of assistants who for the most part hold office for several years. The assistants are encouraged to do original work in the wards and laboratories, have charge, under the professor, of groups of patients, and take a part in the teaching. The original work which they carry out constitutes one of their chief claims to promotion to other posts.

The ideal professor must be a good teacher, and an investigator, and should possess that harmonic influence which stimulates his assistants and others to pursue research. He will suggest lines of investigation, and discourage work on lines which end in blind alleys. He will exercise a general supervision of the labours of the clinique, and will ensure to each worker due credit for the researches which he carries out.[5]

The force of Garrod's quietly articulated but deeply held belief was felt by the Governors of the Hospital, and they moved quickly. Here was an opportunity for Bart's to take a pioneering role in an entirely new venture, the creation of professorial units within a London teaching hospital. They decided, without dissent, to support Garrod's ambitious and revolutionary plan, and on 13 August 1919 Garrod was nominated Director of the Medical Unit, to take office on 1 October. Bart's was thus the first hospital in England to embark on a course that became the model for all teaching hospitals.

In October 1919, the following editorial appeared in the *St Bartholomew's Hospital Journal*:

We are glad to be able to announce an important step in the programme of reconstruction. At the beginning of the Winter Session there will be established in the Hospital a medical and a surgical unit of the "professorial type", with Sir A. E. Garrod and Mr. G. E. Gask respectively as Directors of Clinic. The Directors will be whole-time officers debarred from private practice, and in each case will have the help of an Assistant Director, a first and a second assistant, and two house-physicians or house-surgeons. The units will be provided with wards, clinical laboratories, and out-patient departments.[6]

The Hospital proposed to the University of London that Garrod be given the title of Professor of Medicine; but within three months, and before the proposal could be acted upon, an event occurred that would launch Garrod on a new phase of his career. Osler had died on 20 October 1919, and Oxford, still struggling to recover from the effects of the First World War, was determined to fill the position of Regius Professor of Medicine as quickly as possible.

On 9 January 1920 Garrod received a handwritten letter from Arthur Thomson, Professor of Anatomy at Oxford University:

In complete agreement with my colleagues Dreyer [Georges Dreyer, Professor of Pathology], Sherrington [Charles Sherrington, Professor of Physiology], and Gunn [J. A. Gunn, Professor of Pharmacology], I write you to ascertain whether you would be prepared to accept the Regius professorship here if invited by the Prime Minister.[7]

According to *Munk's Roll*, Garrod was not the only candidate for the Regius Professorship.[8] Sir Wilmot Herringham, Garrod's senior colleague at Bart's, was first approached, but Herringham, who was sixty-four, decided to stay in London. Garrod was not given much time to think it over. Thomson continued: 'We hope you will give the matter your favourable consideration and that you will send me here a telegram on Monday with your decision.'[9]

It was not an easy decision to make. Garrod had great affection for his old hospital, and although he was no longer personally in the forefront of medical research, he had been given the opportunity to set up an experiment in medical education and research for which he had long worked and in which he deeply believed. Moreover, he had already asked Francis Fraser to serve as Assistant Director of the new unit, and George Graham, his old student who had spent so much time with Garrod in Malta, had enthusiastically agreed to join the medical unit as his First Assistant. On the other hand, it was a great honor to be invited to succeed Osler, who had become a legend in his own time. Although Garrod and Osler were not close friends, they had known each other for more than twenty years. Garrod admired Osler greatly, but Osler's outlook on the future of medicine and the need for medical research was very different from his own.

To return to Oxford was ultimately an opportunity that Garrod could not turn down. He saw it, according to Fraser, who succeeded him as head of the new unit at Bart's, as a unique opportunity to 'sow seed on soil uncontaminated by considerations of career and utility'.[10] With some lingering misgivings, for he wondered whether he might be too old for the job at the age of sixty-two, Garrod sent Thomson a telegram saying that he would accept the invitation to become Regius Professor of Medicine if it was offered to him. On 21 February 1920 Garrod received an official letter from 10 Downing Street indicating that the Prime Minister, Lloyd George, was prepared to submit his name to the King for the appointment to the Regius Chair, but before doing so needed to know whether Garrod would accept.[11] Garrod now had no doubts and replied affirmatively.

On Monday, 1 March 1920 a notice appeared in *The Times*:

We learn that Sir A. E. Garrod, recently appointed Director of the Clinical Unit at St. Bartholomew's Hospital, has been recommended for appointment to the Regius Professorship of Medicine at Oxford University, in succession to the late Sir William Osler.

Sir A. E. Garrod is 62 years old, and a Lieutenant-Colonel of the R.A.M.C. Besides his work at St. Bartholomew's Hospital he acts as Consulting Physician to the Hospital for Sick Children, Great Ormond Street. He was educated at Marlborough and Christ Church, Oxford. Throughout the war he served in the Mediterranean as Temporary Colonel A.M.S., was twice mentioned in dispatches, and was made a K.C.M.G. in 1918.[12]

The King acted promptly. On 2 March, Garrod received a letter from Ernest Evans, the Prime Minister's Secretary:

the King has been pleased to approve of your appointment to be Regius Professor of Medicine in the University of Oxford.[13]

On Wednesday, 3 March 1920, *The Times* announced:

The King has been pleased to approve of the appointment of Sir Archibald Edward Garrod, K.C.M.G, M.D., to be Regius Professor of Medicine in the University of Oxford, in the room of the late Sir William Osler, Bt., M.D.[14]

Oxford wasted no time welcoming Garrod. Christ Church immediately elected him a member of the College and to a Studentship. In accepting, Garrod wrote to the Dean on 11 March 1920:

My sincere affection for the House [Christ Church] renders the prospect specially attractive to me — I am conscious that I owe to my undergraduate life there, and the intercourse with men of many different tastes and interests, that wider outlook which makes life so much more interesting.[15]

His letter of acceptance was written on notepaper edged in black, for the loss of his three sons weighed constantly on his mind.

In 1915, Garrod had been invited by the London Hospital to deliver the Schorstein Lecture, but the war and the duties it imposed had precluded his accepting. On 20 February 1920 the Hospital reissued the invitation, and Garrod delivered a lecture on pancreatic disease. The lecture, though scholarly, was not one of his best. The terms of the lectureship required that it be published, and this required additional work at a time when his mind was occupied with his move to Oxford.

Although Garrod's colleagues at Bart's deeply regretted his departure, they wished him well. There had never been a Regius Professor at Oxford from Bart's, and it was a source of pride that Garrod had been selected.[16] There were many farewell parties. On 20 July Garrod was entertained at dinner at the fashionable Café Royal by his former students and colleagues. Despite their regret that he was leaving after an association with Bart's that had lasted forty years, it was a jubilant occasion. Many toasts were drunk, and the evening culminated in the presentation of a silver vase decorated with the two signs of Aesculapius — surmounted by the Cock of Vigilance, and carrying on its side the Serpents of Wisdom. There were also three coats of arms, those of the City, Bart's, and Sir Archibald Garrod. The base bore a wreath of laurel, symbolizing merit, and roses, for affection. The rim carried the inscription:

I was wrought for Sir Archibald Garrod by desire of his late House-Physicians and Assistants at St. Bartholomew's Hospital.[17]

Garrod received congratulatory letters and good wishes from many of the most respected physicians of the day. The Regius Professor of Physic at Cambridge, Clifford Allbutt, wrote:

I know that many good men were alternatively to be considered but it seemed to me so very desirable that for a typically academic post one should be selected who (like Lauder Brunton for example) had distinctly scientific works. Clinical work alone, however good, may be rewarded elsewhere.[18]

Garrod especially cherished the congratulatory letter written by Norah Powell, the sister (head nurse) in charge of Hope Ward, to which Garrod frequently admitted his patients. She had resigned her position upon Garrod's departure for Oxford, and her letter is a touching example of the affection that he instilled in those who worked for him:

Dear Chief,

Thank you so very much for your charming letter, which gives me great pleasure each time I read it over, and which I will always treasure; it is good to know my work has really been of value to you, for every day I have had the privilege of being part of "The Firm" has been a joy to me and I think you will understand that it was impossible for me to continue in a ward where every single thing reminds me that I shall never be able to have the joy again.

Indeed, one cannot imagine you doing no work, surely you will continue in Oxford to inspire the young men with enthusiasm. There should be even greater scope than here, and it is a rare and wonderful gift to be able so to do.

Thanking you from the bottom of my heart for your kindness to me through the years, and with sincere wishes for your happiness in your new life.[19]

Of Garrod's legacies to Bart's, the one which had given him the greatest pleasure was the establishment of the first medical professorial unit in the University of London. The creation of the unit had been largely due to his quiet but persistent urging, coupled with the solid administrative backing of his friend and senior clinical colleague Sir Wilmot Herringham, who was an original member of the University Grants Committee, and who had held the influential position of Vice-Chancellor of the University of London. Garrod left Bart's with mixed feelings, but the call to Oxford was irresistible, and made more pleasant by his colleagues in London insisting that he remain a Governor of the Hospital and Consulting Physician.[20]

When Garrod returned to Oxford as Regius Professor of Medicine he was almost sixty-three years old. He would remain as Regius for seven more years before offering his resignation to the Crown, which enabled him to resign his professorship and take early retirement.[21] The Garrods left London and 5A Montague Mansions, where they had been living after the war, and were off house-hunting in Oxford, and eventually found a comfortable Victorian house at 155 Banbury Road. A new career was about to begin.[22]

The Regius Professor

THE history of medicine at Oxford stretches back to the tenth century. The University's first Regius Professor of Medicine was appointed by Henry VIII in 1546, and remains the senior and most significant chair in medicine at Oxford. Despite the centuries-old existence of the chair, however, Oxford had to await the appointment of Henry W. Acland in 1857 for the natural sciences to get fully under way. Although Acland believed that all medical students should be well grounded in the sciences, and by helping establish the Honour School of Natural Science did much to promote this objective, his contributions to the resurgence of the medical school have been greatly overemphasized. Perhaps his most significant achievement, as far as the medical school was concerned, was to recruit John Burdon-Sanderson, who was generally regarded as one of the leading medical scientists and physiologists of the day, as the first Wayneflete Professor of Physiology. Though Burdon-Sanderson, who later became Regius Professor of Medicine, certainly played a key role in reshaping the medical sciences at Oxford, it was during William Osler's tenure as Regius Professor that the Departments of Pathology and Pharmacology became professorial. Moreover, it is clearly to Osler's credit that he helped persuade Charles Sherrington to leave Liverpool to take up the Wayneflete Chair of Physiology.[1] The standing of Oxford in the medical sciences continued to grow with each successive Regius Professor; after Burdon-Sanderson, William Osler was appointed, and it was Osler whom Garrod succeeded. Although Garrod had not been personally engaged in clinical research for many years, he was an enthusiastic, effective, and articulate spokesman for the need to cultivate a scientific approach to human disease.

In 1920, soon after Garrod arrived in Oxford, J. B. S. Haldane, who was then a Fellow of New College teaching physiology, delivered a paper to the Oxford University Junior Science Club on 'Some recent work on heredity'.[2] Garrod never referred to this paper, and it may very well have escaped his attention, as the Club was small and composed of undergraduates. The paper is historically important, however, for in it Haldane raises the possibility that genes may control enzymes:

> The precise nature of their [genes'] activity is uncertain, but in some cases we have very strong evidence that they produce definite quantities of enzymes, and that the members of a series of multiple allelomorphs produce the same enzyme in different quantities.

Haldane does not refer to Garrod in this short paper, and it is a matter of speculation whether he was aware of Garrod's *Inborn errors of metabolism* at the time. Haldane claimed in 1957 that he had known 'Bateson, from 1919';[3] but, although it seems quite likely that Bateson had told Haldane about Garrod's work, Haldane did not refer to it until much later. The following year, Haldane left Oxford to join Gowland Hopkins at Cambridge, where he became immersed in the physico-chemical aspects of enzyme function.

Garrod was delighted to be returning to his old University. His new position would, he realized, give him a better and more visible platform from which to preach his vision of a 'Higher Medicine'.[4] When Acland resigned from the medical staff in 1879 to devote himself fully to education, the University decided that the Regius Professor's appointment as consulting Physician at the Radcliffe Infirmary could be held *ex officio*. Thus there was no canonical necessity for the Regius Professor to concern himself with clinical work. Burdon-Sanderson, who had studied under the great French physiologist Claude Bernard, had also avoided clinical responsibilities in favor of pursuing his own physiological interests. When Osler took over in 1905, he wrote with some sadness to his friend William S. Thayer, at the Johns Hopkins Hospital: 'The clinical facilities are good as far as they go, but there has not been any attempt to foster practical work.'[5] Although initially discouraged by the apparent lack of clinical emphasis, Osler was not downhearted for long, and was soon expressing to Thayer his pleasure in his clinical work.

The clinical responsibilities for both Osler and Garrod included teaching medicine at the undergraduate and postgraduate levels. Osler frequently used his rounds, which as many as 20 people would attend, to illustrate some particular clinical point or to elicit a specific physical sign. Garrod's rounds, on the other hand, tended to be much smaller, for he never achieved Osler's apostolic clinical following. There was another difference between Osler and Garrod. Unlike Osler, Garrod looked at disease from a biological rather than a practical point of view. He was not concerned with the minutiae of treatment, preferring instead to reflect on the complex genetic and environmental influences that had brought his patient to the hospital with clinically overt disease. Why had one patient come down with influenza or a peptic ulcer while others had not? Garrod's rounds soared with scientific concepts and new ways to look at medical problems, while Osler demonstrated his superb clinical and teaching skills, and preached a down-to-earth pragmatism.

In Garrod's time, the Department of Medicine was inconveniently housed in the cramped quarters of the University Museum, the brainchild of Henry Acland. The Department was almost totally bereft of space, and consisted of the Regius Professor's rooms, his small private laboratory, and, later, an office for the Dean. A small amount of additional laboratory space and lecture rooms were available at the Radcliffe, but they were poorly equipped and wholly inadequate for conducting scientific research. This was a challenge for Garrod, because although he no longer wished to pursue his own research, he believed in its critical importance, and was determined to redress this lack of scientific emphasis.

The opportunity to limit his clinical responsibilities was attractive to Garrod, who saw the Oxford School of Medicine, with its great strength in basic science, as having special academic advantages compared to the London schools. When Garrod approached the Rockefeller Foundation in an attempt to obtain funds for the newly appointed Professor of Biochemistry, Rudolph A. Peters, he felt that it was necessary to outline for the foundation his view of the special role that Oxford could play in medical education:

The place of the Oxford School in the scheme of Medical Education in this country is not to be judged from the numbers of its students. Our Medical Graduates form but a very small fraction of the members of the profession; but a large proportion of the teachers in other Universities and Schools, and of the staff of the great Hospitals are Oxford men. I can recall a time when all the in-patient physicians at St Bartholomew's, and another period when all those at St. Thomas's, were graduates of this University [a possible exaggeration].[6]

Garrod also asserted that what made Oxford medical graduates special was their superior scientific training:

The training which they have received in Anatomy, Physiology and Biochemistry, and especially the practical and experimental work, which forms an essential part of their training, enables them to approach the study of clinical Medicine from a scientific standpoint, and to realise that physiology is the true basis of medicine. The teaching which they receive from the Lecturer in Clinical Physiology serves to emphasize that connexion.

Those who stay here for an additional year to study Pathology and Pharmacology, gain still further advantages, and the Oxford School affords an excellent training not only for men who aspire to the higher walks of practice, but also for those who intend to devote their lives to laboratory work; as witness the many and important Physiological, Pathological and Pharmacological posts held by our graduates in other Universities and Medical Schools.

The fact that there is no advanced Clinical School in Oxford is due to the way in which medical education has developed among us, not under the auspices of the Universities, but around the great hospitals [the London teaching hospitals], to which the students were first admitted as apprentices of individual members of the staffs. If this system has its disadvantages, as it undoubtedly has, it may claim to be the parent of the English system of clinical teaching, at the bedside, and to limited groups of clerks or dressers, who are the lineal descendants of the old apprentices.[7]

Although the Medical School was strong in the physical sciences, Garrod and his colleagues did not believe it would be wise for Oxford to develop, at that time, into a complete clinical school. The Dean of the Faculty of Medicine, E. W. Ainley Walker, enlarged on this in a note to the Rockefeller Foundation:

It would be futile if our medical students after being trained in an extended course, and to a high level in Medical Science, should attempt to complete their Clinical studies in Oxford. They therefore all pass on to the great Clinical Schools in London (occasionally elsewhere), returning to Oxford for their Final Examinations.[8]

The Radcliffe Infirmary was, *de facto*, a provincial hospital of some 200 beds, and did not serve a sufficiently large population to provide Oxford students with adequate clinical experience. This suited Garrod well, for he much preferred to develop the Radcliffe as a center for postgraduate studies where clinical research could be undertaken along the lines he had planned for the medical professorial unit at Bart's.

Some of us at least believe that it would be highly desirable to establish in Oxford an organised, post-graduate clinic, in which some of our graduates, and other picked men, might pursue the study of scientific clinical *Medicine*, under the direction of a Professor chosen as specially fitted for such a task, and aided by a group of trained assistants; but, in the present circumstances, we see little prospect of a speedy realization of this project.[9]

This was quite unlike the role that Osler had carved out for himself at Oxford. Science had not been his first priority, and although he had recruited a faculty that was very strong in the basic sciences, Osler himself emphasized in his rounds the art of clinical medicine and the cultivation of broad scholarship in the history of medicine and the humanities.

Garrod's administrative accomplishments during his tenure as Regius Professor were significant. He had scarcely settled into 155 Banbury Road when he found himself elected to all the major University committees, including the Medical Faculty Council, the Board of the Faculty of Medicine, and the General Board of Faculties, responsible for the overall academic administration of the University. One of his first administrative actions was to appoint the first Dean of the Medical School. This departure from long tradition — a departure that was not received with universal approval — was made more palatable because his candidate, Ainley Walker, was an able and well-liked pathologist. Ainley Walker, who was elected Dean on 22 November 1920, proved to be a man upon whom Garrod could rely implicitly.

Garrod also served on the time-consuming Hebdomadal Council, the governing body of the University, which met every week during term to discuss the academic affairs of the University, or to put it more formally, to deliberate on all matters relating to the maintenance of the privileges and liberties of the University. The records of the Hebdomadal Council show that Garrod was extremely conscientious in fulfilling his adminis-trative responsibilities, faithfully serving, often as chairman, on numer-ous subcommittees. He scrupulously edited draft reports, clearing up possible ambiguities and revealing his own careful analysis and under-standing of administrative detail.[10]

On 21 May 1922 Garrod successfully argued before the Council that medical students 'shall be exempted from examination in chemistry, physics, and biology, provided that [they have] passed these subjects while attending previous College or University, and that the examina-tions for these subjects have been approved by the General Medical Council.'[11]

Not all Garrod's University responsibilities were related to medicine. The President of Magdalen College called the attention of the Hebdomadal Council to the inadequacy of the title 'Fielding Herbarium' as a description of the Department of Botany. The council promptly rewarded the President of Magdalen for his suggestion by instructing him to look into the matter. An additional member was deemed necessary, Garrod was elected, and in due time the Department of Botany was so named. Garrod was also asked to consider matters relating to the future of the Radcliffe Observership, the astronomical arm of the University. This assignment must have reminded him of his student days.

Soon after his arrival, Garrod was elected a Delegate of the Oxford University Press. In this latter capacity, Garrod, like Osler before him, voted on whether manuscripts submitted to the Press were worthy of carrying the Oxford imprimatur. He appeared to enjoy these re-sponsibilities, but seldom made specific suggestions. The Press had printed both editions of his *Inborn errors of metabolism*, and later, on the eve of his retirement, he asked that the delegates consider publishing a book on diathesis. Oxford University Press were not particularly enthusiastic about publishing the essay, but because Garrod was a Delegate of the Press and because they had published *Inborn errors of metabolism*, they reluctantly agreed.[12]

On 27 May 1920 Garrod's old friend from his Malta days, Sir Themistocles Zammit, received the degree of Doctor of Letters, *honoris causa*. It is not hard to imagine that Garrod, in a spirit of friendly academic reciprocity, played a significant role in the award, for in 1916 Garrod had received an honorary MD degree from the University of Malta, with Zammit's enthusiastic support.

In May 1922 the University of Padua celebrated its seventh centenary, and Oxford University was invited to send two delegates. Garrod was selected to attend, and received an honorary degree. Of all the honorary degrees that he collected during his lifetime, none gave him greater pleasure than this. After all, Garrod's own hero William Harvey had graduated from Padua with high honors as Doctor of Medicine in 1502 before returning to England to work at Bart's.

Garrod used his influential position as Regius to ensure that his scientific approach to medicine would be shared by his new colleagues. Shortly after his appointment to the Medical Faculty Council, that body, chaired by Garrod, drafted new admission requirements for candidates for the graduate (qualifying) degree of Bachelor of Medicine (BM). Furthermore, it was decided that Garrod, as Regius Professor of Medicine, would have to approve the subject of a candidate's doctoral dissertation, which should be 'connected with the science or art of Medicine in further refining the topic', and which would allow those few physicians who aspired to a higher medical degree to attach the letters DM (Doctor of Medicine) after their name.[13] Garrod revised the draft report to emphasize the importance of scientific research as well as general excellence:

The subject must be previously approved by the Regius Professor of Medicine [and] shall contain at least an original contribution to the science and art of medicine, and that the subject must be based upon personal observation or research.[14]

A meticulous examiner of doctoral theses, Garrod insisted on clarity of expression, as well as scientific accuracy, as the following excerpt from one of his comments illustrates:

Mr. Gilles has submitted in his thesis the results of much hard work. It is based upon his own observations of a very large number of cases of Malaria, mostly as might be expected of the Subtertian type. His experiments recall clearly what we saw in Malta during the war, in cases from the Vardar Valley. As a literary production the thesis leaves much to be desired. 'Cases' 'die' and 'recover' and the English often falls short of what might be wished, moreover he has not sufficiently corrected errors of the typist, obviously without medical training.[15]

Garrod wanted the Faculty of Medicine to earn the academic respect of all the faculties of the University, and the minutes of the Hebdomadal Council provide recurring evidence of his insistence on scientific and scholarly excellence. He clearly believed that he would have a more lasting effect on medicine in Oxford by stressing that the future of medicine lay in applying scientific principles to human disease, rather than by pursuing the more Oslerian role of exemplary bedside teaching. Above all else, Garrod wanted Oxford to encourage in every way possible the furtherance of medical science. Both Garrod and Osler were interested in medical history, but while Osler was a highly distinguished and internationally recognized medical historian and bibliophile, Garrod was more concerned with the evolution of scientific ideas, particularly his belief in the importance of biochemical individuality.

Perhaps influenced by his appointment to the Council of Somerville College, an Oxford college for women, as well as by his daughter Dorothy's academic ambitions, Garrod proposed in May 1924 that women undergraduates be admitted as candidates for the medical degree and that they be eligible to apply for Radcliffe Travelling Fellowships, which enabled successful applicants to travel abroad. Garrod was persuasive, and his recommendation was adopted without any recorded demurral.

Dorothy Garrod is said to have embarked on an academic career in archaeology partly as a result of a book that her father brought home. As a Delegate of the Press, he was entitled to a copy of all the books published by the Press. Knowing of Dorothy's nascent interest in archaeology, he artfully left a book on the subject on a table in the living room. We also know that Dorothy visited Malta with her parents about 1920, met Zammit again at that time, and was immensely stimulated by him.

Dorothy had obtained her Diploma in Anthropology, with distinction, at Oxford in 1922, and Garrod undoubtedly remembered that the award of a traveling scholarship had given her the opportunity to work with the formidable Abbé Henri Breuil, the distinguished French paleonto-logist at the Institut de Paléontologie Humain in Paris. This work inspired her to proceed to an Oxford B.Sc., which she passed in 1924, again with distinction. Garrod enjoyed having Dorothy nearby and watching the successful launching of her career, which included a visit to Malta to investigate the figurines uncovered at the Tarxien temples. Her enthusiasm was a heartening recompense for the loss of his three academically promising sons. Dorothy's scholarship steadily matured, and she became an outstanding prehistoric archaeologist (Fig. 31).

The Oxford calendar of events indicates that Garrod, like Osler before him, delivered two series of teaching rounds. Every Monday at 2 p.m., he gave a clinical lecture to undergraduates, and on Tuesdays at 2.30 p.m. he lectured to the postgraduates. He also followed Osler's lead in conducting an informal teaching round every Sunday at 11 a.m. Occasionally, he agreed to read the lesson at Christ Church on Sunday — readings that were well remembered for their inaudibility. Whether these evangelical diversions interfered with his regularly scheduled Sunday rounds is not known.

All those who recall Garrod's years as Regius Professor agree that conducting general clinical rounds did not give either him or his audience particular pleasure; but the shrewder medical students soon discovered that they could greatly enliven the rounds by selecting patients suffering from rare diseases such as hematoporphyrinuria or

Fig. 31 Dorothy Annie Elizabeth Garrod *circa* 1965, when she became a Commander of the British Empire in recognition of her outstanding achievements as an archaeologist.

hemophilia. Rare diseases not only gave Garrod the opportunity to discuss his favorite subject, inherited metabolic disorders, but also allowed him to discourse more generally on the need to cultivate a scientific spirit — a necessity if the students were ever to have a glimpse of the 'Higher Medicine'.

Although a fair clinician, Garrod never pretended to be a medical oracle, dropping clinical pearls before rapt medical students, nor did he expand on Hippocrates, or Galen, or the art of medicine, as had Osler. Neither did Garrod possess Osler's easy charm, cheerfulness, and irrepressible confidence. But while the students and junior faculty did not hold Garrod in clinical awe, and certainly did not regard him as a 'Great Physician' in the Oslerian mode, they found him kindly and avuncular and, if they listened, always stimulating. Although it had been many years since he had worked at the bench, he was exceedingly popular with the scientists working in the University Museum, which housed the basic sciences. Garrod appreciated the importance of basic science, and felt at home discussing their problems. Osler, though beloved for his clinical enthusiasm, good humour, and sense of fun, 'seemed to the young physicians to be rather out of date, particularly with his enthusiasm for examining blood, in unfixed and unstained preparations'.[16]

Garrod was never happier than when he was discussing what was new in science, the role that science would play in arriving at a better understanding of human disease and the advancement of medicine, and the critical importance of biochemical individuality in manifestations of clinical disease. In a letter to the author, Ronald Macbeth, formerly consultant Ear, Nose, and Throat Surgeon at the Radcliffe Infirmary, gives an evocative picture of Garrod:

When I knew him around 1925–26 he was a gentle, rather tired, old man. He was broad-shouldered and woolly in appearance — large moustache and a mop of grey hair going a bit thin in the middle.

In those days Oxford did not have a full clinical school, but the 4th year people still up and doing "bugs and drugs" (Pathology and Pharmacology) used to go to ward-rounds, demonstrations etc. at the Radcliffe. Garrod, though having no care of beds, would give a clinical talk on such patients as interested him at (I think) 5 p.m. on Wednesdays. It was always a scholarly presentation, involving usually a medical history and the by-ways of human metabolism.

His personality must have been the antithesis of that of Osler, whom he succeeded ... and the Osler ghost must have been a little tiresome to Garrod. This was rather a shame, because even Osler's most ardent admirers would never call him a scientist. Garrod's outlook was scientific.

Garrod continued one of Osler's ploys, and with great success. Under his Chairmanship there would be a ward round for doctors at 11 a.m. on Sundays. The medical and surgical staff members took it in turn to produce the patients, and we students were allowed to tag along. It was most stimulating, because in the discussions no holds were barred, and we were at an early stage disabused of the idea that the doctor is always right. Over this session Garrod presided with tact.[17]

Students are always alert to perceived idiosyncrasies, and the students at Oxford were no exception. An often-recited student ditty, emphasizing the peculiarities of the physicians and surgeons of the time, had the refrain:

> You really must listen to the Regius,
>   The Regius, the Regius,
>   The inimitable lectures of the Regius.
> But for ailments rare and curious,
> Such as Haematoporphyrinurias,
> You certainly must listen to the Regius.
>   The Regius, the Regius,
>   You must never miss a lecture by the Regius.[18]

Clearly, the students regarded Garrod as their good-natured, professor who liked to discuss 'ailments rare and curious'; but there is no evidence to suggest that they considered him a profound thinker.

Garrod's broad knowledge of biochemistry enabled him to take a role in many of the branches of science pursued within the University. He was elected, in both 1922 and 1923, chairman of a committee that supervised the practical work in organic chemistry, and he served as an examiner in the subject — an astonishing accolade for a physician who had no formal association with organic chemistry since leaving Oxford in 1890. He also served as acting head of the Department of Biochemistry after Benjamin Moore unexpectedly died in 1922 from cardiac failure following an attack of flu, and before Rudolph A. Peters was recruited from Gowland Hopkins's Department of Biochemistry at Cambridge in 1923.

In addition to inculcating a scientific spirit into medicine, Garrod made another notable and tangible contribution to academic excellence during his years at Oxford by obtaining funds from the Rockefeller Foundation to enable the Department of Biochemistry to become firmly established and financially secure. Hans Krebs, Whitely Professor of Biochemistry, who succeeded Peters at Oxford, has already given an excellent account of Garrod's role in gaining Rockefeller Foundation support for the Department of Biochemistry;[19] and, more recently, Margery Ord and Lloyd Stocken have written an admirable account of the history and activities of the Department.[20] Rockefeller Foundation records contain memoranda that further demonstrate Garrod's pivotal role in seeing that Biochemistry played a major part in the University. His good friend Hopkins was forging ahead, building up the Biochemistry Department at Cambridge, and Garrod was eager to catch up.

On 7 July 1922, at a meeting of the Executive Committee of the Rockefeller Foundation, Richard W. Pearce, Director of Education, was authorized to make a survey of medical education in Europe. As part of his assignment, Pearce paid a visit to Oxford on 2 February 1923; there he was welcomed by Garrod, who introduced him to James A. Gunn, Professor of Pharmacology, and E. W. Ainley Walker, Dean of the Medical School. During the morning they toured the laboratories of Gunn and Charles Sherrington, as well as those of Professor of Pathology Georges Dreyer, who had been recruited from Copenhagen. After lunch Garrod took Pearce on a walking tour of Oxford, making sure that the Radcliffe Infirmary was a conspicuous part of the itinerary.[21]

Pearce was impressed by a medical faculty that included, in addition to Garrod, Sherrington in Physiology and the young, recently appointed Peters in Biochemistry. Garrod expressed his hope that the Foundation would bestow a sizeable gift upon the University to support the newly created Department of Biochemistry. Pearce was clearly receptive; in an internal memorandum the following day, he wrote: 'The scientific side of the work at Oxford has developed rapidly in recent years ... and one is tempted to consider rounding out their plan by filling the gap in Physiological Chemistry.'[22]

Garrod and Sherrington had persuaded Peters to come to Oxford from Cambridge, where he had been an assistant to Hopkins. Although Hopkins would have been delighted to keep Peters in Cambridge longer, he recognized that an opportunity to go to Oxford at the age of thirty-four as a Professor was not to be missed, and he urged Peters to accept at once. In accepting Garrod's offer, Peters gained a nice title — the Whitely Professor of Biochemistry — but not much else. If Peters were to launch biochemistry at Oxford in the way he wanted, he would need support to pay the salaries of a small staff and to provide some essential laboratory equipment to get started. Following Pearce's visit, Garrod wrote to Peters in Cambridge and asked him what he would need. Peters replied promptly with an estimate of £141 000; £60 000 for the building, £14 000 for maintenance, and an additional £67 000 for general support.[23]

Fortunately, the Rockefeller Foundation proved to be the fairy god-mother that Garrod had hoped for. By the autumn of 1923 Pearce had persuaded the Foundation that there was an opportunity to build an exciting new scientific department at Oxford, and returned to England to pursue what he now called his Oxford initiative. He met Garrod on 23 September and was presented with Peters's estimate. In the Rockefeller Foundation's frugal style, Pearce asked for some reduction, pointing out that Garrod had anticipated private funding that had not yet materialized. In view of this uncertainty, Pearce recommended to the Foundation that they delay definitive action until 1 October, when Oxford's new Vice-Chancellor, Joseph Wells, was due to take office. When Garrod and Pearce lunched together in October, Garrod provided him with a revised estimate of £106 400. With the firm promise of government money, and with some additional private support that Garrod had up his sleeve, the Foundation agreed to provide Oxford with £75 000 toward a building endowment.[24]

Matters moved swiftly from then on. Garrod asked Pearce if the Foundation would provide funds immediately, in order to stimulate other potential donors. On 6 December, Pearce told Garrod, with whom he had developed a warm and collegial relationship, that his efforts to promote physiological chemistry at Oxford had been approved by the Foundation, and that it would provide the sum of £75 000 forthwith. It was a nice Christmas present for Garrod, Peters, and the University. Garrod was in Cambridge examining when he heard the news. He had planned, after completing his responsibilities at Cambridge, to go to Torquay, an increasingly fashionable seaside resort on the south coast, for a few days' rest. He was so relieved and elated by the news that, instead of going directly to Torquay, he stopped at his club in London to write an informal letter of thanks to Pearce. In his official letter to the Foundation, Garrod later wrote: 'As Regius Professor of Medicine, I am most deeply grateful, and know that the other teachers of the Faculty share my gratitude.'[25]

There was still the vexing problem of getting the University to guarantee an annual sum of £250 in perpetuity, requested by the Foundation, but Garrod was justifiably optimistic that it could be done. Indeed, he could be well pleased with his entrepreneurial efforts: his young protégé Peters was now well set up, with the academic future of the Department in his hands. Garrod's quiet but persuasive advocacy on behalf of the University and the Medical School had finally borne fruit.

In a memorandum to Pearce in January 1923, Garrod had expressed his opinion on what the role of Regius Professor of Medicine should be:

Nowadays the Regius Professor being in no way overburdened with routine teaching, can perform useful functions as a spokesman for medicine, and for science generally, on the administrative bodies of the University, and has immense opportunities of influence if he chooses to avail himself of them.[26]

One of Garrod's minor obligations as Regius Professor of Medicine was to become the Master of the almshouse at Ewelme, nestling at the foot of the Chiltern hills in Oxfordshire. This responsibility, ordained by James I in 1617 for the care of about 24 elderly people, was neither onerous nor time-consuming, and Garrod clearly enjoyed his occasional overnight visits to the lovely village of Ewelme, as well as handily augmenting his modest salary.

For the most part, Garrod remained busy, a respected elder statesman fulfilling his administrative duties and modest teaching responsibilities. In 1923 he joined the Medical Research Council, and remained on its Board until 1928. Unfortunately, the records of the Council were lost during the Second World War; had they survived, they would have provided additional insight into the extent of Garrod's influence on medical research in England. Garrod was a popular external examiner, and enjoyed the intellectual stimulation that accompanied examining with colleagues at other universities. In addition to supervising all the medical examinations at Oxford, Garrod was also an examiner in medicine at most of the universities awarding medical degrees, as well as for those taking the membership examination for the Royal College of Physicians. As an examiner, he had the reputation of being acute and keenly probing, but always kindly.

Garrod's medical colleagues frequently turned to him to deliver distinguished named lectureships and to address a variety of medical gatherings, and he used these occasions to portray the intellectual excitement that lay ahead for those who would approach medicine from a scientific viewpoint. When the second edition of *Inborn errors of metabolism*, with a printing of 1000 copies, was published in 1923, fifteen years after his Croonian Lectures, Garrod wrote to Hopkins asking if he might dedicate the second edition to him. On 17 July 1922 Hopkins wrote, clearly delighted:

The Dedication is a source of great pride for me! In fact your most friendly act has really given me more sincere pleasure, I think, than anything I can recall during my scientific career. You see 'Inborn Errors' has really meant very much to me in the past, as a source of inspiration for myself, and as a potent aid for awakening enthusiasm in one's students. To be thus associated with the book is therefore something really *big*.[27]

The book had not changed greatly, but Garrod had added two chapters to it. One was on hematoporphyrinuria, and one was on congenital steatorrhea, both inherited diseases of metabolism that had captured his interest. The reviewers of the second edition were more generous than their predecessors. The journal *Nature* provided a full-page review of Jack C. Drummond, the physiologist, of 'this most valuable monograph'.[28] The reviewer for the *Lancet* applauded the work, but clearly regarded the subject as a byway in medicine. The diseases discussed were so rare 'that they do not enter into practical medicine'. Clearly, the reviewer admired Garrod, and remarked that the work would be read 'by all medical men, because of the fascinating way in which the author deals with the subject'.[29] There is no hint that the reviewer is aware of Garrod's emphasis on biochemical individuality.

Garrod devoted 26 pages to porphyria in the second edition. The chemistry of the porphyrin molecule was beginning to be understood, principally by Hans Fischer and his school in Germany, and Garrod, with his familiarity with the German language, had no difficulty in keeping up with a contemporary understanding of the chemistry of this complex tetrapyrrole.

In 1921 Leonard Mackey and Garrod had encountered a six-year-old boy who passed red urine.[30] This feature was noted at the first micturition, thus establishing in the eyes of Garrod that this was an authentic case of recessively inherited congenital porphyria. A remarkable feature of the patient was the deep pink coloration of his milk teeth, as well as skin that was exquisitely sensitive to light.

Ever since his early spectroscopic studies, Garrod had been interested in colored urine, and for the next decade porphyria occupied much of his attention. He no longer worked in the laboratory himself, and had persuaded E. N. Allcott, a chemist in Oxford's Department of Organic Chemistry, to undertake detailed chemical studies of the boy's urine and feces. Because porphyria is a rare disease, Garrod inquired particularly about parental consanguinity; but this was excluded. Garrod retained a deep interest in the patient, and subsequently presented his case at a meeting of the Association of Physicians at Birmingham. When Garrod wrote his last paper, in 1936, the patient was nineteen years old, his teeth had turned brown, and his skin had become mummy-like and scarred from frequent infections of light-induced bullae of the skin. His spleen and liver, impalpable when he was seen in 1921, were now greatly enlarged.[31]

Garrod is clearly describing the disease now more commonly known as congenital erythropoietic porphyria. His reliance on spectroscopy to detect minute quantities of porphyrin has not been superseded. The amniotic fluid surrounding affected infants contains large amounts of porphyrin, and confirms Garrod's strong hunch that the disease is present at birth. Garrod's conjecture that the disease is exceptionally rare remains true today. The failure of Garrod to find a history of parental consanguinity in most of his patients may be explained, partially, by the fact that the disease is very heterogeneous.

The significance of Garrod's work on inborn errors of metabolism was finally becoming more widely accepted; but the diseases that he studied were still regarded as rare and recondite, and therefore of little relevance to the medical profession as a whole. Garrod himself believed that he had merely discovered the tip of a very large evolutionary iceberg, and that there were many more inborn errors of metabolism to be discovered. But, he pointed out, if such errors did not cause overt disease, they would have to advertise themselves in some other, very evident way if they were not to be missed: 'One man in 20 000 whose urine of twenty-four hours habitually contained a few grammes of aspartic acid might well be overlooked'.[32]

Although Garrod realized that metabolic diseases, particularly if common, were likely to interest his various audiences, he always returned to what he regarded as an overriding biological truth — one even more important than the inborn errors that caused disease — namely, the phenomenon of biochemical individuality, both between species and within species.

This way of looking at human disease was, almost certainly, sown in Garrod's mind when he read Carl H. Huppert's paper on the maintenance of the characteristics of the species.[33] But whether it originated with Huppert or not, there is no question that Garrod saw inborn errors of metabolism as part of a spectrum of normal variation. Some variations were trivial, even harmless, while others were so severe that they caused death. The complex chemical variation of man, the result of diverse evolutionary pressures over millions of years, was the biological background against which Garrod viewed all disease processes. In this regard he was not only quite unlike any other physician of his day, but his views were probably incomprehensible to them. Garrod wrote clearly, and his ideas were not bizarre; but they went unheeded. It was not only physicians who did not grasp the profound significance of his work, however; geneticists and biochemists were equally unimpressed.

It is important to recognize that the first chapter of the 1909 edition of *Inborn errors of metabolism* provided strong evidence of Garrod's remarkable insights into evolution. Indeed, he took pains to place the inborn errors in an evolutionary context:

The delicate ultra-chemical methods which the researches of recent years have brought to light, such as the precipitin test, reveal differences still more subtle, and teach the lesson that the members of each individual species are built up of their own specific proteins, which resemble each other the more closely the more nearly the species are allied.

... Nor can it be supposed that the diversity of chemical structure and process stops at the boundary of the species, and that within that boundary, which has no real finality, rigid uniformity reigns. Such a conception is at variance with any evolutionary conception of the nature and origin of species. The existence of chemical individuality follows of necessity from that of chemical specificity, but we should expect the differences between individuals to be still more subtle and difficult of detection. ... Even those idiosyncracies with regard to drugs and articles of food which are summed up in the proverbial saying that what is one man's meat is another man's poison presumably have a chemical basis.[34]

Garrod stressed in his essay that even structural malformations such as harelip, cleft palate, and spina bifida are probably 'malformations by arrest', and result from failure of some step or other in the series of chemical changes which constitute metabolism. He was fascinated by the behavioral characteristics of the waltzing mouse: '... their bizarre dance is merely the functional manifestation of an inborn and hereditary malformation of the semi-circular canals'.[35]

In May 1923 Garrod was asked to deliver the Linacre Lecture at Cambridge, a lecture in the gift of the Master and Fellows of St John's College. He chose the tantalizing title 'Glimpses of the Higher Medicine'.[36] His theme, however, was not new; he spoke, as he had so often before, on the role of science in clinical medicine. But it was a message worth repeating, for the relevance of science to clinical medicine was receiving little more than token acknowledgement among the clinicians of the day.

In his Linacre Lecture, Garrod asked his audience to consider clinical medicine in an evolutionary perspective:

If it be granted that the individual members of a species may vary from the normal of the species in chemical structure and chemical behaviour, it is obvious that such variations or mutations are capable of being perpetuated by natural selection; and not a few biologists of the present day assign to chemical structure and functions a most important share in the evolution of species.

In a sentence that seems to herald the present biotechnological revolution, Garrod, perhaps prompted by Georges Dreyer, urged that the natural defenses of the body should be isolated and administered as therapeutic agents:

When the various protective agents shall have been isolated, so that we can give them in pure form, and freed from extraneous substances which hamper their use, we shall be far better equipped than we are at present for the fight with disease.

Later he enunciated a general law of seminal importance:

whereas the intruding agents [of disease] are the actual cause of disease, the reaction of the patient's tissues shapes the clinical picture.[37]

Although the usefulness of clinical biochemistry in the diagnosis and evaluation of clinical disease was gaining increased recognition, any interest beyond that was difficult to discern. This lack of curiosity troubled Garrod, and at every opportunity he would ask his audience, as he did in his Linacre Lecture, to look beneath the superficial categorization of disease and to think about the disease itself. What were the environmental and inborn components of disease and susceptibility to disease? How did they affect the metabolism of the organism? Garrod did not dwell on the inborn errors of metabolism in his lecture — except to point out that diseases inherited in a simple Mendelian fashion did not have to be present at birth. He regarded Friedrich's ataxia, myotonia congenita, and muscular dystrophy as diseases for which 'it is difficult to escape the conclusion that although these maladies are not congenital, their underlying causes are inborn peculiarities'.[38] In analyzing the basis for his belief that biochemical individuality was fundamental, he again refers to Huppert's 'masterly rectorial address of 1895'.[39]

One of Garrod's preoccupations was a desire to understand cellular metabolism and its biochemical perturbations in clinical disease. He was fascinated by control mechanisms; some he felt he understood, at least in outline, and he gave examples from the field of endocrinology; but others, equally important, left him bemused: 'In what way do enzymes control, within physiological variations, the blood pressure, the temperature, even an individual's height?' Garrod was thinking out loud, puzzled, as we still are today, by the delicate balance of the myriad metabolic activities of normal cells, each controlled by a single gene. He thought it amazing that the processes did not go awry more often. What did it mean, biologically, for an individual to be susceptible to a disease or to respond anomalously to a drug?

He ended his lecture with a sentence as succinct as it is profound: 'In short, diathesis is only another name for chemical individuality.' This recurring theme, elaborated in detail but unchanged in principle, formed the title of his essay, *The inborn factors in disease*, published eight years later, in 1931.

In 1924, Garrod delivered the Harveian Oration at the Royal College of Physicians, the most prestigious of all College lectureships, and chose as his title 'The debt of science to medicine'.[40] Garrod took his audience through the evolution of scientific thought, emphasizing how many of the early scientists had been trained as physicians. The great names of science — Thomas Young, Luigi Galvani, Hermann Boerhaave, and William Wollaston — were all physicians. He told his audience that 'Physics and chemistry are the fundamental sciences upon which physiology and medicine rest', and quoted, perhaps for the first time, the now well-known letter in which Harvey, six weeks before his death, replied to a Dutch physician who had asked for his opinion on an unusual pathological specimen:

It is even so — Nature is nowhere accustomed more openly to display her secret mysteries than in cases where she shows traces of her workings apart from the beaten path; nor is there any better way to advance the proper practice of medicine than to give our minds to the discovery of the usual law of Nature, by careful investigation of cases of rarer forms of disease. For it has been found, in almost all things, that what they contain of useful or applicable is hardly perceived unless we are deprived of them, or they become deranged in some way.

Garrod went on:

These words, as true to-day as when they were written, are full of encouragement for those of us for whom the study of Nature's experiments and mistakes has a special attraction. The structural malformation, or the hereditary and inborn departure from the normal of metabolism, although unimportant from the practical standpoint, may throw a ray of light into some dark place of embryology or biochemistry; and not a few of the rare maladies, such as chloroma, polycythaemia vera, sulphaemoglobinaemia, and the disease of which Bence–Jones albuminuria is a sign, offer fascinating and still unsolved problems of physiology and pathology.

Obviously clinical medicine presents immense fields of scientific research, and those who cultivate them have the added satisfaction of knowing that every advance of medical science will, sooner or later, bring in its train some forward movement of the healing art.[41]

Garrod returned to the general theme of science and medicine when he was invited to give the opening address to the medical students at the winter session of the Westminster Hospital in 1926.[42] He was careful not to belittle the need to study the sick at the bedside, and to be skilled in the art of therapeutics. A wise physician should cultivate 'tact, resourcefulness, courage, and prudence'; he must have 'patience with fads, consideration for his patients and their friends, sympathy with suffering, and gentleness of touch and voice'. Garrod was well aware that these qualities were central to becoming a good physician; but he preferred to discuss medicine on a more sophisticated level. In this lecture, he again returned to the overriding and crucial truth of biochemical individuality:

Seeing that no two individuals are exactly alike either in structure or in chemistry, sickness does not conform to any single model; each individual case calls for careful observation. Owing, as I believe, to their chemical individuality different human beings differ widely in their liability to individual maladies, and to some extent in the signs and symptoms which they exhibit.

He emphasized to his fellow physicians their medical ignorance, and the eternal need for more research:

> What do we really know of the true nature of the leukaemias, despite the many differential blood counts of recent years? These carry us to a point, and seem to be unable to take us further. A new road needs to be blazed out, by the collaboration of the laboratory and the ward.[43]

Thomas Archibald Malloch, a Canadian physician whose father was a good friend of Osler, became part of the medical team that attended Osler during his last illness. After Osler died, Malloch worked for a short while with Garrod, and they kept up a lively and informed correspondence after he left. In October 1926 Garrod wrote from Melton describing how he and Laura had attended the meeting of the British Association the previous August. His mood is relaxed:

> We came here on Aug. 12th, much later than usual, after a very strenuous Brit. Ass. time. It went off very well. Dorothy had a star turn with her Gibraltar skull. Her book came out about the same time, but the skull rather puts it in the background. She has been having a holiday here with us, and is now on her way back to Gib. to finish the site, which should take about 3 months more. We hope to go out in the Christmas vac. I expect you have seen about the Neanderthal skull which she found. We went into a garage in Ipswich the other day to have something done to the car, which took about an hour. When we came back the workman who had done the job took me to the window to pay; and when I gave my name he said: 'Is the finder of the skull a relation of yours by any chance!' I replied that it was my daughter who was in here with me an hour ago; 'Well I'm blowed' was his reply. Such is fame![44]

Although Garrod was invited by C. C. Hurst, a poultry geneticist who had become interested in human genetics, to a meeting at the London School of Economics in July 1931 to discuss the creation of a British Council for Research in Human Genetics, 'a subject which interests me greatly',[45] Garrod was planning a trip to Norway and did not attend, nor did he ever join, the Council. He also never joined the Eugenic Education Society, inspired by Galton and founded in 1907, nor did he join or attend any of the meetings of the Genetical Society. Genes were nebulous; human disease, on the other hand, was real and insistent. Unseen beads of unknown chemical structure distributed linearly along the chromosomes were philosophical abstractions, and Garrod admitted that he could not follow the work of Lancelot Hogben and other geneticists. Consanguinity was important in his understanding of the inborn errors of metabolism; but the science of genetics seemed to have added little over the years. Even when Hopkins founded the Biochemical Society, Garrod did not join. In contrast, he continued to enjoy the meetings of the Physiological Society, a group and a discipline in which he felt more at home.

It should not be surprising that Garrod was not swept off his feet by genetics. Clinical chemistry and the principle of biochemical individuality had continued to advance understanding of disease; but genetics seemed to hold little clinical promise. Population genetics seemed hardly relevant since the chromosomes of man had not been usefully identified or counted. Although Garrod was fully supportive of undertaking research in human genetics, it is unlikely that organizations designed to effect 'eugenic improvement' would have impressed him. He was then seventy-four, and had little motivation to become immersed in the polemical issues surrounding the emerging eugenic movement.

11

Creativity in Retirement

WHEN Garrod retired as Regius Professor of Medicine in December 1927, there was the usual round of farewell dinners and other celebratory occasions (Fig. 32). He had enjoyed the scholarly atmosphere at Oxford, but he was now ready to leave, and on 20 December he chaired

Fig. 32. Archibald Garrod at the time of his retirement as Regius in 1927.

a meeting of the Hebdomadal Council for the last time. When the Council next met, in February 1928, his successor, E. Farquhar Buzzard, was Regius, and a new era had begun.[1]

In January 1927 Garrod was elected by the Committee on Secondary Education of the Hebdomadal Council to serve a three-year term as Governor of Ipswich School in Suffolk.[2] Garrod was undoubtedly pleased with this assignment; the duties were light, and this was the school that had educated his father and other Garrods of his father's generation. His daughter, Dorothy, was at Cambridge pursuing an increasingly successful career in prehistoric archaeology, and Garrod's old friend Gowland Hopkins remained Professor of Biochemistry there. Garrod decided to take up permanent residence at Wilford Lodge, the home in Melton, Suffolk, inherited from his grandfather. Melton was within striking distance of Cambridge, and sufficiently close to London for Garrod to attend meetings of the Royal College of Physicians and its College Club, as well as the Royal Society, of which he was still a Vice-President.

Garrod had been for some years a staunch supporter of the Oxford Museum of the History of Science, and before leaving he donated various memorabilia, including a traveling thermometer originally owned by his grandfather, a Gregorian telescope, and a cupping glass that had belonged to his father. His brother Alfred's sphygmograph, as well as a handmade bent-tube thermometer, Garrod also gave to the Museum. In addition, he presented two monaural stethoscopes that had served him well for much of his early professional life.[3]

Although these gifts might indicate that Garrod's career in medicine was over, his interest in biochemistry and inborn errors of metabolism was undiminished. He never ceased to ponder the significance of biochemical individuality and its relevance for human disease. Although he had proclaimed it for twenty-five years, Garrod's insight that biochemical individuality was a crucial determinant of disease had been neither understood nor appreciated by his colleagues.

An invitation to deliver the Huxley Lecture during November 1927 at Charing Cross Hospital enabled him to link the principle of chemical individuality with the half-forgotten doctrine of diathesis, and to emphasize once again the importance of constitutional factors in the incidence and shaping of maladies.[4]

In this lecture, named in honor of Thomas Huxley, Garrod looked at diathesis from an evolutionary as well as a medical aspect:

It is with unfavourable deviations that the student of diathesis is concerned, but if there were no beneficial ones there would be no evolutionary advance; progress could go no further.

With his usual meticulous care, he introduced his subject by paying tribute to those who had discussed constitutionality or diathesis in the past. He drew a distinction between the statistical anthropometric approach, originally proposed by Francis Galton and now practised in the clinic of George Draper in the United States, and his own. Draper and his pupils approached diathesis from 'four different directions: the anatomical, the physiological, the psychological, and the immunological'. Garrod did not deride this globally ambitious approach; but he preferred to examine specific examples of inborn errors of metabolism and see what light they might throw on the general problem.[5]

He discussed what he called the borderland between structural and functional anomalies, and focused on 'tissue defects', referring to a number of dominantly inherited disorders of bone such as fragilitas ossium, muscular dystrophy, and disorders of red blood cells, including hemolytic disease. He differed from those who regarded gout as a primary defect in renal function, preferring instead to consider it an abnormality of purine metabolism. Today, the prevailing view is that in most patients decreased excretion is the most prominent factor, while in some patients there is an overproduction of uric acid. Garrod admitted that his earlier work had not paid sufficient attention to the development of arthritis in alkaptonuria, which 'was long thought to entail no evil effects'. He believed that albinism was probably caused by the lack of an enzyme which brings about the formation of melanin, but he also stated, with his usual objectivity, that 'a much more striking example [of an inherited disease] is afforded by hemophilia, a typical sex-linked recessive, the essential cause of which appears to be absence of a coagulative factor'.

Garrod's Huxley Lecture is of particular historical importance because it formed the basis for his essay on the inborn factors in disease, published four years later. With characteristic modesty, he closed with a warning:

We need to beware lest such phrases as "error of metabolism", used without real justification, may become convenient cloaks for our ignorance.[6]

He concluded with a statement that underscores the profundity of his biological approach to diathesis, foreshadowing by more than half a century the contemporary explosion in biochemical polymorphisms:

Again, I have said little of the converse of diathesis, the benign mutations which favour the individual in the struggle for existence. Difficult as it is to detect an error of the body chemistry by its evil effects, which may be long postponed, it must needs be more difficult to detect those which are harmless or have only good effects. Yet we can see how such favourable mutations have been utilized in the building up of the defensive mechanisms which are ready to be brought into play to counteract a poison introduced, and which confer upon some individuals and on the members of certain families an inborn immunity, partial or complete, from various infectious maladies.

A year earlier, in November 1926, Arthur Hurst, Physician to Guy's Hospital and a long-time acquaintance of Garrod, had delivered an address to the Ulster Medical Society on 'The constitutional factor in disease'.[7] It is instructive to compare the two lectures, published within six months of each other. Hurst's lecture appears vague and somewhat confused, in sharp contrast to the electric clarity of Garrod's. Believing that a 'short' stomach secreted an abundance of acid, Hurst had coined the term 'the hypersthenic gastric diathesis' to indicate, as he believed, that those who inherited 'short stomachs' were particularly likely to suffer from duodenal ulceration. He believed that pernicious anemia and achylia gastrica were also associated with a gastric diathesis, and concurred with Draper that gallstones were constitutionally determined.

Hurst's capacious umbrella of 'constitutional disease' sheltered asthma, epilepsy, and migraine from sharp scientific scrutiny. He came closer to the mark, however, when he stated that hypertension and hypotension were variations of the normal, and that hypertension was probably an inherited condition; but he rather spoilt it by suggesting that hypertensive patients 'react in a special way to toxaemia and to mental and physical fatigue'.[8] Although he referred to Garrod and inborn errors of metabolism in his introduction, and agreed that such entities exist, Hurst did not think it 'possible to improve upon Draper's conception of constitution'.

The editorial that accompanied Hurst's lecture provides a useful contemporary view of 'constitutional factors'.[9] The term had come into vogue in the nineteenth century, and had been incorporated into the teaching of such luminaries as Thomas Addison and Jonathan Hutchinson. However, when bacteriology burst on the scene towards the end of the century, discussion of constitutional factors in the causation of disease gave way to a rash of enthusiasm for microbes. Forty years later, the concept was revived; Hurst in England and Draper in America were among its most vocal advocates.

Draper in his constitution clinic in New York used anthropometric methods in an attempt to identify individuals who were susceptible to certain diseases. Hurst was more attracted to a physiological approach. Although he had always admired Garrod, and clearly recognized Garrod's pioneering role in understanding the inborn errors of metabolism, Hurst, like his contemporaries, failed to appreciate that what Garrod had done was to translate the vague notion of constitutionality into the precise concept of biochemical individuality.

Garrod may have been puzzled that the profession did not appreciate his biological and biochemical viewpoint; but he was not one to thrust his opinions on others. He had none of Bateson's aggressiveness; like Darwin, he needed a Huxley to preach to the unconverted. In 1929, two years after his Huxley Lecture, in an elegant address to the Osler Club Garrod dwelt on Osler's magnetic personality, whose 'human sympathy and unselfishness promoted in every way the brotherhood of their profession'. Discussing the personality of those engaged in research, he was unwittingly portraying himself:

The great investigator may plough a lonely furrow, and have little personal contact with his fellow men, and yet his writings and the results which he obtains, his discoveries, may change the whole course of human thought.[10]

At the end of 1927, at the request of the Rockefeller Foundation, Garrod wrote a piece on the teaching of clinical medicine in England and his reflections after a career in medical education and research. It had been eight years since Garrod had persuaded Bart's to create the first medical unit:

It is too soon as yet to pronounce upon the success or failure of this experiment, and especially for one to express an opinion who is rather an advocate than an unbiased critic. Not a few of those who are slow to welcome an innovation and who cling to established customs, those indeed who never felt the want of such new departures, will give unfavourable replies. Those who expect too much from the introduction of new wine into old bottles are also liable to be disappointed and to look for quicker returns than can be secured in the short time during which the system has been used.

He concluded:

It seems little likely nor it is desirable that the unit system will replace the established methods, but those who favored its introduction firmly believe that it is destined to play an important part in the advancement of medical education and the progress of medical science, in this country.[11]

Despite his hesitancy, Garrod has clearly been vindicated: sixty-five years later, the 'unit system,' modified by the passage of time, is going strong. Special research units were to become common in the United States, and thrived at the Johns Hopkins Hospital, the Thorndike Laboratory of Boston City Hospital, and the Massachusetts General Hospital.

Garrod was invited in June 1928 to open the Courtauld Institute of Biochemistry attached to the Middlesex Hospital in London. He used this opportunity to explain the need to understand the relationship of nature and nurture to human disease, and to express optimism that this need was being slowly recognized:

Of recent years, under the spell of the advances of bacteriology and protozoology, we have tended to lay all the stress upon the invading malady and to pay too little attention to the reaction of the organism invaded, but there are signs that the pendulum is returning from the limit of its swing, and in its modern dress the revived doctrine of diathesis will rest largely upon a chemical basis.[12]

Garrod emphasized again the crucial importance of looking at man and his diseases in chemical terms:

Medicine may be looked upon from several distinct standpoints, but, in importance, the chemical standpoint is second to none, as is becoming more clearly recognized from year to year. Biochemistry is not merely a useful preliminary subject of study for the medical man, but is part of the very essence of his science, and, through his science, of his art.[13]

In an intriguing sentence he suggested that mutations may be of chemical origin: 'It would seem that there is a chemical basis for those departures from type which are styled mutations, and I for one believe that the liabilities of certain individuals to, or their immunity from, certain maladies — what may be called their diatheses — have chemical origins.' It is frustrating not to know what was going through Garrod's mind when he wrote those words; a year earlier, H. J. Müller had shown the influence of radiation on mutations. Garrod almost certainly knew about this work, and, although he does not refer to it explicitly in his lecture, he does refer, without attribution, to the relationship between X-rays and mutation in *The inborn factors in disease* three years later.

In 1927 a small group of general physicians who were principally interested in diseases of children formed the British Paediatric Association. At its inauguration, three physicians, Thomas Barlow, Humphrey Rolleston, and Archibald Garrod, were elected to honorary membership. Although Garrod had not been a practising pediatrician for many years, his work on inborn errors of metabolism had begun with the children he had seen at the Hospital for Sick Children at Great Ormond Street. When he retired from his honorary consultantship at Great Ormond Street a one-time colleague, Frederic J. Poynton, recalled his impressions. Like many others, Poynton remembered Garrod for his courtly, reserved manner, and commented that 'he was far too learned to be a confident physician and was only really happy with a test tube in his hand and attending patients with rare disorders of metabolism'.[14]

Garrod cherished his years at Great Ormond Street, and his election to the brotherhood of pediatricians pleased him greatly. In the same year he was elected an honorary member of both the Association of American Physicians and the British Association of Dermatology and Syphilology, the latter probably at the urging of his colleague at Great

Ormond Street, Edward A. Cockayne, whose interest in genetics had been fired by Garrod's work on inborn errors of metabolism.

The honors and awards heaped on Garrod at the time of his retirement as Regius Professor were evidence of the esteem in which he was held by his medical colleagues. But in no way did they acknowledge or recognize the far-reaching consequences his discoveries would have for the understanding and treatment of human disease. Garrod had made no great, direct contribution to therapy, and he was not revered as a great teacher or humanist. Despite his many years as a caring physician, his inner contentment came from his contemplation of biochemical individuality and the way it would ultimately revolutionize medicine. That revolution, however, would lack a following for many years; even today it is a view not universally accepted.

Garrod delivered his last major address, the annual oration of the Medical Society of London, in 1928.[15] Founded in 1773 and still thriving, the Medical Society of London had not joined the various other medical organizations that together formed the Royal Society of Medicine in 1907. Unlike the Royal Society of Medicine, the Medical Society of London had as one of its stated missions the provision of regular lectures.

Garrod's father had served as president of the Medical Society of London, and Archibald had been a member for many years, delivering his Lettsomian Lecture on glycosuria in 1912. The Society's rooms at 11 Chandos Street were almost next door to Number 10, where the Garrods had lived before he was called to Oxford, and the Garrods frequently attended the Society's functions.[16]

The lecture, much of which was historical, contained discussions of a number of rare diseases, including porphyria, Gaucher's disease, dystrophia myotonica, intermittent hydrarthrosis, and Fröhlich's syndrome. Apologetically, Garrod admitted that:

I should have preferred, had it been possible to say nothing this evening about inborn errors of metabolism, a subject with which I fear to weary my colleagues; but to omit all reference to alcaptonuria would be to forgo some of the strongest arguments which can be brought forward in support of my thesis.[17]

In early January 1928 Sir Dyce Duckworth had died, and Garrod was asked to write his obituary for the hospital reports. While it was perhaps his filial duty to praise his old chief, Garrod genuinely admired Duckworth's clinical skills. Moreover, during the formative years of Garrod's clinical training at Bart's, Duckworth had been one of his enthusiastic supporters. Their careers, however, could not have been more different. Duckworth's view that materia medica was a recondite and lost art rather than a science and his belief that gout represented a disorder whose primary disturbance was in the medulla oblongata of the brain were among many opinions to which Garrod could not subscribe. Nevertheless, Garrod's appreciation of Duckworth as a clinician was sincere, and the obituary was characteristically generous and kindly.[18]

Garrod displayed a rare example of his sense of the whimsical in a letter to the *St Bartholomew's Hospital Journal* in 1929, in which he referred to the obituary of J. F. Bullar, another friend and contemporary:

He kept a hedgehog, as a pet, in the residents' quarters, and repute had it that it lived in his bath. This involved cockroach hunts at night, and the Dispensary was a rich covert — in the book which had to be signed by residents visiting the Dispensary at night such entries as the following: Name of patient — Timothy Hedgehog; disease — starvation; drug required — *Blatta domestica*.[19]

In 1931 Garrod published his second classic work, *The inborn factors in disease*[20] (Fig. 33); 1250 copies were published. Oxford University Press had not been particularly enthusiastic about publishing this volume, but agreed to do so because of its long association with Garrod, both as an author and as a Delegate of the Press.[21] The concern of Oxford University Press was misplaced: *The inborn factors in disease* proved to be Garrod's crowning intellectual achievement.

Garrod had always been struck by the variability of human response to environmental perturbations, such as infectious agents, foreign antigens, and drugs. Mankind, Garrod insisted, did not possess a single, evolutionarily near-optimal genotype, but one with an array of variations, ultimately expressed as the biochemical uniqueness of the individual.

In *The inborn factors in disease* Garrod gave specific examples of the relevance of biochemical individuality to disease, and devoted several pages to 'The inborn factors in infective disease'. Cogently, he observed that tuberculosis could not occur without close contact with the tubercle bacillus, and malaria did not exist without the presence of the plasmodial vector.

Nevertheless it must never be forgotten that it is not only in causing predisposition that internal factors are concerned; but also, that upon the patient's constitution depends the form which the morbid syndrome assumes.

Various forms of hypersensitivity, often with a familial aggregation, were selected by Garrod as examples of idiosyncratic reactions of the host to external factors. The explanation for human allergies was to be found 'in the chemical individuality of the subject, and those of his cells'. Garrod gave a weird, perhaps unique, example of idiosyncracy drawn from his personal experience:

THE INBORN FACTORS
IN DISEASE

AN ESSAY

BY

ARCHIBALD E. GARROD, K.C.M.G.

D.M., LL.D., F.R.C.P., F.R.S.

Consulting Physician to St. Bartholomew's
Hospital, and to the Hospital for Sick Children ;
sometime Regius Professor of Medicine in
the University of Oxford

OXFORD
AT THE CLARENDON PRESS
1931

Fig. 33 Garrod's second classic work, *The inborn factors in disease.*

The present writer, who has not, to his knowledge, any excessive sensitivity to any other flower or plant, and who does not suffer from hay-fever, is unable to stay in a room with the flowers of the orange-coloured buddleia without experiencing general discomfort and breathlessness. The purple buddleia, on the other hand, has no such effect.

Garrod had long been aware of idiosyncratic reactions to therapeutic agents. Struck by the variation in human response to various drugs, he pointed out that many drugs after they enter the body are metabolized by enzymes in the liver. He speculated that individual variations in the liver enzymes might influence the response of patients to a particular drug. His pharmacological colleague at Oxford, J. A. Gunn, with whom he discussed these idiosyncratic reactions, added the perceptive suggestion that it was likely that ferments in the liver:

are usually occupied in dealing with ordinary products of metabolism, rather than that they lie in wait in the tissues in the hope that some day their appropriate alkaloid may come along.[22]

Today's burgeoning field of pharmacogenetics would have given Garrod quiet pleasure, for he was its pioneering spirit.

Garrod sent a copy of *The inborn factors in disease* to his friend Sir Thomas Barlow, then aged eighty-seven, a former President of the Royal College of Physicians and Garrod's senior colleague at Great Ormond Street. Barlow, a talented investigator and clinician, was the first to show that infantile scurvy and adult scurvy were identical. In an appreciative and insightful note, he wrote:

I always thought your work on chemical inheritances was more epoch making than we realised at the time. Issue some more of your reflections, recollections and anticipations, I beg of you, even if they don't all weld yet. Their chemical base is surely a field not completely ploughed yet and there are not many people qualified to plough it.[23]

Although *The inborn factors in disease* was reviewed on both sides of the Atlantic,[24-27] it did not create any particular stir. The unidentified reviewer for the *Lancet* suggested that it would be easier and more rewarding for physicians to study Draper's classification of diathesis than to try to understand Garrod's complex genetical exposition.[28] In March 1932 the Historical Section of the Royal Society of Medicine held a meeting, clearly in Garrod's honor, to discuss the history of 'the introduction of biochemistry in medicine'.[29] Garrod took exception to the title, preferring to think of biochemistry as originating a branch of medicine. In discussion, Gowland Hopkins, President of the Royal Society, emphasized the debt that biochemistry owed to medicine. Deciding who deserved the title of Father of Biochemistry, Hopkins said, was difficult, but he 'had decided in the end that so complex a subject deserved two: Justus Liebig and Sir Archibald Garrod'.[30] Although Hopkins was undoubtedly influenced by the nature of the occasion, his admiration for Garrod's contributions was not feigned. It was clearly a joyous educational affair, full of wit and, as Garrod put it, 'a thread of satire'.

Two years before the publication of *The inborn factors in disease* Garrod had decided to move from Melton to Cambridge. The initial pleasures of country life had begun to pall; he felt intellectually isolated at Wilford Lodge, and moving to Cambridge would bring him more into the academic mainstream. The Hopkinses were delighted to have the Garrods in Cambridge, and Archibald was a frequent visitor to Hopkins's Department of Biochemistry. Garrod got on well with the faculty and students, and they, in turn, found him immensely stimulating. To his great delight, Garrod was elected a Fellow of Emmanuel College, to which Hopkins also belonged, and he enjoyed his dining privileges there.

The Garrod family friendship with the Hopkinses had been strengthened when Jacquetta Hopkins, Gowland Hopkins's younger daughter, decided to embark on a career in archaeology. She was eighteen years younger than Dorothy, but over the years they developed a strong personal and professional friendship. She had become an Associate at Newnham College in 1951, and, with Dorothy, excavated in France and Palestine. A distinguished author and archaeologist herself, she married fellow archaeologist Christopher Hawkes in 1933, and subsequently, in 1953, J. B. Priestley.

Jacquetta Hopkins's early childhood recollections of Archibald Garrod are somewhat at variance with his avuncular reputation. She remembers him as ruddy-faced and inclined to be short-tempered. Short and stumpy, with bushy eyebrows, he was a contrast to his wife, whom she recalls as large-framed, heavy-lipped, and often unsmiling. Jacquetta's mother always remembered that when the Garrods came to dine it was not a relaxed evening.

In Cambridge, the Garrods were able to see more of Dorothy, who held a Research Fellowship at Newnham College when she was not away on one of her digs in Kurdistan, Gibraltar, or Palestine. Cambridge was to recognize Dorothy's scholarship some years later in 1939 when she was appointed to the Disney Chair of Archaeology, the first woman to be elected to a professorship at the University. The Garrods adjusted happily to the less demanding tempo of their lives in Cambridge, and once again found time for pleasures from their early married life, such as cruising the Mediterranean and visiting the Greek Islands. They also revisited Malta, to which they had first taken Dorothy in 1916. The Garrods enjoyed these regular cruising holidays, but after the first few years of retirement, Archibald's disabilities increased, and they were no longer able to travel.

In 1933, Garrod was invited back to Bart's to speak to the preclinical students.[31] He described the strides made in physical chemistry that were becoming increasingly relevant to medicine, particularly advances in the study of enzymes and their actions.[32] It was not necessary, he reassured his audience, that they should all become biochemists, 'but you do need to know something of the biochemical approach to disease. You need to retain a mental picture of the chemical processes of which the body is the seat.'

Garrod was now seventy-six years old, and although his mind was as acute as ever, he was becoming physically more frail and inclined to be irritable. His eyesight began to fail; he had developed retinal macular degeneration, a progressive disease for which there is no treatment, and which leads ultimately to blindness. As his eyesight worsened, he depended on Laura to read to him, though he was able to correspond with close colleagues. The Garrods continued to visit the Hopkins household, but Archibald no longer felt able to visit the Department of Biochemistry. Garrod had always been fascinated by puzzles, and in the last years of his life the solution of the daily crossword puzzle was, according to his daughter, a family event; as his vision failed, he became skillful at solving anagrams in his head.

In May 1934, Garrod received, as a member of the Harveian Society, a copy of the Buckston Browne Prize Essay. The essay, 'The influence of disease on heredity', had been written by Lionel S. Penrose, who subsequently held the Galton Chair of Eugenics at University College, London, and whose distinguished contributions to human genetics were to become widely recognized.[33] In his essay, Penrose had given credit to Fritz Lenz, in Germany, for being the first to draw attention to the relationship of consanguinity to recessive characters.[34] Garrod, in his reply, gently reminded Penrose that it had been Bateson who 'saw daylight' and had first suggested that recessive inheritance would explain the high rate of consanguinity in alkaptonuria. Garrod went on to refer Penrose to his 1903 paper,[35] written in German, where:

[I] discussed the matter at some length. Previously, I believe the Mendelian theory had not been applied to human characters. As the idea is definitely of British origin one is reluctant to see any credit to which it may be entitled assigned to Germany.[36]

Garrod wanted no credit for himself:

> I am not suggesting that you should do anything about it, nor writing with any sense of grievance. Indeed, in Hogben's paper on alcaptonuria in the *Edin. Royal Soc. Proceedings*, and in his "Nature and Nurture"[37] he has, I think, given me more credit than I am entitled to, seeing that it was Bateson who saw daylight. I approached the subject not from the side of genetics primarily, but from that of alcaptonuria.

This was not the first time Garrod had indicated that the 'new genetics' was beyond him. In a letter to E. A. Cockayne dated 22 March 1933, Garrod had written:

> I find myself quite out of my depth in the new Mendelism of Hogben and Haldane. Have you seen Hogben's paper on the genetic basis of alcaptonuria in the *Proceedings of the Royal Society of Edinburgh*? I hesitate to mention it, because he refers too kindly to a thirty year old paper of mine, but if you don't know it you will find it interesting. It is curious to look back to the old Bateson–Weldon controversies and the position of Mendelism today.[38]

When Penrose received Garrod's letter, he replied immediately, apologizing for 'my ignorance of many parts of the subject on which I wrote and I am very glad you have drawn my attention to the early work on alcaptonuria'.[39] In his defense, Penrose correctly claimed that 'I think I am right in stating that Lenz (1919) was the first person to arrive at the exact mathematical expression connecting the frequency of a recessively determined condition and the incidence of consanguinity in the parents.'[40] There was clearly no ill feeling between the younger and the older man, and in August 1935, Penrose again wrote to Garrod,[41] drawing his attention to a recent paper by Asbjörn Fölling, a Norwegian, who had just described for the first time what is now known as phenylketonuria.[42] News of this inborn error of metabolism had not reached Garrod, and he was fascinated to learn about it, immediately writing to Fölling for a reprint. In his reply, Fölling wrote:

> I am very proud, that the author of "Inborn Errors of Metabolism" likes to have a reprint of every paper upon phenylpyruvic imbecility. For the time being I have collected 16 cases of the disease in this country.[43]

The last honor Garrod received was the Gold Medal of the Royal Society of Medicine. His old friend Robert Hutchison, who years ago had come down from Scotland to work with him at Great Ormond Street, was now President of the Society, and had been eagerly looking forward to presenting him with the medal. In July 1935 Garrod received Hutchison's invitation, which indicated that the Gold Medal would be presented at a dinner during the Annual Meeting of the

Society the following February. Garrod was deeply touched by the honor, but wrote back that he feared his health might not allow him to come to London at such an inhospitable time of the year.

Although the Garrods were leading an increasingly quiet life in Cambridge, Archibald's mind remained active, and he continued to keep up with the literature, particularly with his latest interest, the inherited porphyrias. In 1936 the editors of the *Quarterly Journal of Medicine* agreed to Garrod's request that he should write a Postscript for the *Journal* on three cases of congenital porphyrinuria that had been previously described. One of the patients had been discussed in 1922[44] and again in 1926. In commenting on the hereditary aspects of the disease, Garrod was clearly puzzled:

Familial occurrence is undoubted; in not a few instances two or more members of a sibship born of normal parents have been affected. Some subjects have been children of consanguineous parentage, but one would expect a larger proportion with such parentage if the anomaly be a Mendelian recessive. In such, the greater the rarity of the condition, the higher the proportion of consanguineous parents to be expected. There is no doubt that congenital porphyria is extremely rare — rarer even than alcaptonuria.[45]

When the proofs arrived, Garrod was too sick to make a final revision of the paper. For several years he had suffered from angina pectoris and cardiac asthma, but now the frequency of these attacks had increased and they were incapacitating him. Sadly, time had run out, and he would not live to see the paper published. On Saturday, 28 March 1936 Archibald Edward Garrod suffered a coronary thrombosis and died at his home at 1 Huntington Road. He was seventy-eight.

Garrod's death did not pass unnoticed, but his contributions to medical science, in death as in life, were not fully appreciated. Although the obituary in *The Times* the following Monday occupied two columns, the author barely touched on Garrod's contributions:[46] 'His chief work was concerned with the more recondite chemical problems connected with metabolic changes in disease.' Gowland Hopkins wrote obituaries for the journal *Nature*[47] and, later, for the Royal Society.[48] He also contributed an appreciation in the *British Medical Journal*, in which he noted that on one occasion Garrod had said: 'I fear, Hopkins, I have never been sufficiently interested in the routine treatment of disease.'[49] In the *Lancet*, Hugh Thursfield wrote a three-page tribute to his old friend and pediatric colleague of twenty years.[50] He recalled that at the end of his career, having left Oxford for Cambridge, Garrod had dryly remarked that it was almost as though he were changing Dryden's famous lines:

Cambridge to him a dearer name shall be,
Than his own mother-university,
Thebes did his green, unknowingly youth engage;
He turns to Athens in his riper age.[51]

In the obituary in the *Lancet*, Francis Fraser, Garrod's successor as Director of the Medical Unit at Bart's, characterized Garrod thus:

An able, practical physician when the need arose, patients, as he said himself, did not really interest him, and the complex problem presented by an individual who is ill did not really appeal to him for solution ... His was the mind of the true scientist and often one wondered, and he wondered also, how he had come to be a practising physician. He was almost an onlooker when he applied himself to the daily task of seeing patients and treating them. This daily task was something that had to be done, a price that had to be paid, for the privilege of contact with the numerous intriguing puzzles that he found and set himself to solve. Often he would spend all morning at a window in the ward testing a urine that presented an unusual colour or smell. The morning's round could be left to others ...

Inclined to long silences, with bushy eyebrows and gentle voice, he had a personality that claimed attention ... He could be sarcastic but was never unkind, so when he censured strongly there was good reason for it.[52]

In all the obituaries, including one in the *Journal of the American Medical Association*,[53] *Inborn errors of metabolism* was mentioned, but little reference was made to *The inborn factors in disease*, published only five years before his death, and none to the importance of chemical individuality. The minutes of the Faculty of the Royal College of Physicians and Surgeons of Glasgow emphasized his pioneering role in the creation of the 'unit' system in London medical schools, as well as his influence on clinical pathology.[54] The obituary for the *St Bartholomew's Hospital Journal*[55] was written by the metabolically minded physician, one-time pupil, and long-time friend George Graham, who referred to *The inborn factors in disease* but did not regard it as particularly noteworthy. Perhaps it was too subtle for this London clinician. 'The book is interesting to read, but it has not the fire or importance of *Inborn errors of metabolism*.'

Thus, the significance of Garrod's biological contribution was largely overlooked by the medical community of the time, and he was usually drably and narrowly defined by the press and in his obituaries as a 'chemical pathologist and physician'. In April a memorial service was held at the church of St Bartholomew-the-Less in Smithfield.[56] In addition to Garrod's family and old friends from London, Oxford, and Cambridge, a galaxy of the country's most distinguished physicians came to pay a last tribute to their colleague. None of them realized that

Fig. 34 A stained glass window in the Hall at Christ Church, Oxford, depicting the coat of arms of Garrod. Garrod was admitted as a student in Christ Church in 1876. The background for the panel reads 'Dominus illuminatio mea', 'May God light my way.' By kind permission of the Governing Body of Christ Church, Oxford.

the intellectual concepts of the man whom they mourned would later be regarded as seminal to the development of medicine, biochemistry, and genetics for the remainder of the twentieth century.[57]

A year after Garrod's death, the Medical Council of St Bartholomew's Hospital recommended that his name be affixed to one of the wards of the medical unit. In the event, the name Garrod Ward was given to a female medical ward on the fourth floor of the King George V medical block; in 1988, it became a mixed medical ward.[58] In Oxford, a memorial plaque to Garrod rested in the Medical Science Museum until Henry Harris, Regius Professor of Medicine (1979–92), removed it to the Sir William Dunn School of Pathology, of which he was Director, and where the plaque is now placed. In the Hall at Christ Church, stained glass windows depict, with heraldic decoration, three Regius Professors of Medicine: Henry Acland, William Osler, and Archibald Garrod (Fig. 34).

12

One Gene–One Enzyme

Regardless of when it was first written down on paper, or in what form,
I myself am convinced that the one gene–one enzyme concept was the
product of gradual evolution beginning with Garrod.

— George W. Beadle[1]

WHEN Garrod summarized the inborn errors of metabolism in 1909, his
work was indifferently received by his contemporaries. Put aside, if not
actually forgotten, Garrod's Croonian Lecturers gained remarkably in
currency fifty years later, when George W. Beadle in his Nobel Lecture of
1958 paid an almost over-generous tribute to Garrod's work:

In this long, roundabout way, first in *Drosophila* and then in *Neurospora*, we had
rediscovered what Garrod had seen so clearly so many years before. By now we knew
of his work and were aware that we had added little if anything new in principle. ...
Thus we were able to demonstrate that what Garrod had shown for a few genes and
a few chemical reactions in man was true for many genes and many reactions in
Neurospora.[2]

Beadle's tribute rapidly led to the widespread belief that Garrod was the
originator, if not the actual 'father', of the one gene–one enzyme
hypothesis. This appealing conclusion, as Sapp has pointed out, is an
attractive fable, an oversimplification based on hindsight.[3] Moreover, as
Charles R. Scriver and Barton Childs have emphasized, it is certain that
Garrod could not have concluded that his studies on inborn errors
implied a one-to-one correlation between a gene and an enzyme.[4] It is
somewhat curious, as they also point out, that Garrod appeared to resist
using the word 'gene' in his published work, preferring to use the more
general term 'factor'.

Spurred on by Beadle's tantalizing remarks in his Nobel Lecture, numerous scholars have attempted to trace the historical roots of the one
gene–one enzyme hypothesis, and in all such accounts Garrod is given a
central role.[5] The hypothesis has assumed the characteristics of a
scientific slogan. Explicit articulation of the phrase, if not the concept,
however, was first introduced by Beadle, when he wrote:

I have several times been asked when the one-gene–one-enzyme hypothesis was first proposed and by whom. I have thought about this question many times and have reread a number of papers to see if I could discover the answer. I have not been successful. The first reference in so many words that I know of is in a review of biochemical genetics that I wrote for the 1945 volume of *Chemical Reviews*.[6]

When Beadle delivered his Harvey Lecture in New York, also in 1945, he was quite specific:

Furthermore, for reasons of economy in the evolutionary process, one might expect that with few exceptions the final specificity of a particular enzyme would be imposed by only one gene.[7]

Ascribing a particular date or forum for the articulation of the one gene–one enzyme hypothesis may have received too much emphasis, for it was in the minds of Beadle and Tatum in 1940, but they did not regard the idea as particularly revolutionary. In their paper delivered on 1 January 1941 at the annual meeting of the American Association for the Advancement of Science, they wrote:

... since we are convinced by the accumulating evidence that the specificity of genes is of approximately the same order as that of enzymes, we are strongly biased in favor of the assumption. In this we make no claim to originality, for it has many times been suggested by geneticists that there may be a close relation between genes and enzymes. It is, of course, possible that the immediate products of many genes may be enzymes or their protein components. At the present time, however, the facts at our disposal probably do not justify the elaboration of hypotheses based on this assumption.[8]

Thus, although the embryonic concept of the relationship of genes to enzymes can be inferred from Garrod's papers on alkaptonuria, as well as from some of Bateson's writings,[9] there is general agreement that the establishment of a direct relationship between one gene and one enzyme had to await Beadle and his colleague Edward L. Tatum's experimental work on the mold *Neurospora*, a more tractable and well-disposed experimental organism than either man or *Drosophila*.

Robert Wagner[10] has recently asserted that many authors, in discussing the roots of the one gene–one enzyme concept, have paid insufficient attention to the work of the French geneticist Lucien Cuénot.[11] In 1903 Cuénot published a paper on the relationship of Mendel's law to the inheritance of coat color in mice, in which he discussed 'the capacity of an enzyme to react with a material substance', and thus came close to recognizing a direct relationship between Mendel's units of inheritance and coat color. Cuénot's paper was published in a French zoological journal, and would almost certainly have escaped even Garrod's roving bibliographic eye. Bateson, on the other hand, was well aware of Cuénot's work, and referred to it extensively when he wrote *Mendel's principles of heredity* in 1909 and *Problems of genetics*, published in 1913.[12] J. B. S. Haldane in 1954[13] also gave credit to Cuénot in discussing the origins of biochemical genetics.

Between 1900 and the time he delivered his Croonian Lectures in 1908, Garrod was a busy practitioner of medicine, seeing patients at Great Ormond Street, Bart's, and the West London Hospital, as well as the occasional private patient. Along with some of his younger medical and chemical colleagues, he continued to investigate inborn errors of metabolism whenever they presented in his hospital work.

Meanwhile, the field of genetics was beginning to take shape. Bateson was aggressively assembling plant and animal material that could be used to attack the anti-Mendelians, led by Walter Weldon, Karl Pearson, and other adherents of the ancestrian viewpoint. Apart from his occasional participation in the meetings of the Royal Society of Medicine, probably arranged by Garrod, and his address to the neurologists in 1906,[14] Bateson paid little attention to human heredity.

In 1902, the same year that Garrod published his paper on alkaptonuria, Walter Sutton began his study of chromosomes, which culminated in the publication in 1903 of 'The chromosomes in heredity',[15] a work that clearly established the importance of these structures in the study of heredity. Three years later, Bateson, Punnett, and colleagues[16] provided the first evidence for genetic linkage; but, again, there is no evidence that Garrod was particularly influenced by these advances. Garrod may have learned about Sutton's chromosome theory from Bateson, but it seems equally likely that, when they met, Bateson regaled Garrod with the difficulties he was having with the biometricians rather than attempting to explain the intricacies of emerging chromosomal genetics.

It should be recalled that when Garrod began his work on inborn errors of metabolism, the word 'genetics' did not exist. The word was not coined until 1905, when Bateson, in a much-quoted letter to Adam Sedgwick, commented that a recently announced bequest to Oxford University was intended to 'promote the study of and research in the sciences of vegetable and animal biology'. This phrase appeared cumbersome to Bateson, and he went on to suggest:

If the Quick Fund were used for the foundation of a Professorship relating to Heredity and Variation the best title would, I think, be "The Quick Professorship of the study of Heredity". No single word in common use quite gives this meaning.

Such a word is badly wanted, and if it were desirable to coin one, "GENETICS" might do.[17]

When Bateson published *Mendel's principles of heredity* he devoted an entire chapter to the 'evidence as to Mendelian inheritance in man'.[18] He discussed brachydactyly as a dominantly inherited trait, and color-blindness as an example of an X-linked condition. In discussing recessively inherited disorders, Bateson referred to alkaptonuria, describing it as a disease in which 'the urine is red from the presence of alkapton'. This description must have startled Garrod. Apart from the spectral *faux pas*, which Bateson repeated in his Silliman Lectures,[19] he provided a succinct and persuasive account of the genetics of alkaptonuria, but remained puzzled at the report of one case in which the 'chemical fault' appeared to be inherited in a dominant fashion.

Bateson went on to argue that dominant conditions were caused by the presence of a specific factor, while recessive disorders were due to its absence — the so-called presence–absence hypothesis. He cited albinism as being 'almost certainly due to the absence of at least one of the factors, probably a ferment, which is needed to cause the excretion of pigment'. He regarded alkaptonuria as being 'due to the absence of a certain ferment which has the power of decomposing alkapton'. Bateson's conclusion seems to have been stated with conviction, yet he later expressed some doubt:

In regard to some of these [departures from Mendelian expectation], it is, I think, still open to question whether the transmission is a process comparable with that which we ordinarily designate as Heredity. Some element is obviously handed on from individual to individual, but it seems to me possible that this element or poison is distributed irregularly among the germ-cells, spreading among them by a process which is mechanical, like the spread of an oil-stain in a heap of paper, or of a fungus in a heap of seeds. In the present state of pathological knowledge it is premature to make any suggestion as to the possible nature of such poisons ... [F]rom a study of the inheritance in an ample series of families, I am inclined to suppose that the element transmitted is something apart from the normal organism, and that it is handed on by a process independent of the gametic cell-divisions. In such cases I do not anticipate that any "law" of inheritance can be discovered, for if my view is correct, the process is not heredity in the naturalist's sense at all.[20]

Joshua Lederberg has recently emphasized that neither Garrod nor Bateson appears to have reflected on the functional significance of the normal allele, nor did Bateson ever specifically indicate that he thought the normal allele was responsible for 'normal' metabolism.[21]

During the first decade of the century, when Garrod was consolidating his ideas on inborn errors, the Cambridge mathematician G. H.

Hardy had been stimulated by Punnett to think about Mendelian genetics.[22] At a meeting of the Royal Society of Medicine,[23] with Garrod present, Udny Yule argued that if Mendelian rules obtained, a dominant gene would eventually increase to the point where three-quarters of the population would exhibit the dominant phenotype. Hardy perceived that this notion was self-evidently false, but, pressed by Punnett, he agreed reluctantly to send a note to *Science* setting forth simple algebraic evidence that, given certain assumptions, the ratio of affected individuals (carriers) to normal individuals would remain constant. This led to the birth of the classic Hardy–Weinberg law, which is not simply known as the Hardy law because Wilhelm Weinberg, a German physician, had arrived at a similar conclusion at about the same time.[24]

None of these developments seems to have been noted by Garrod, whose interest continued to veer more and more towards normal and abnormal metabolism and further and further away from formal genetics. In Garrod's correspondence with Bateson between 1902 and 1908 there is nothing to suggest any preoccupation with the new genetics. Garrod was quite content to remain on the sidelines as an interested, if remote, observer. Genetics as a discipline never held him in thrall.

The war years prevented Garrod from pursuing biochemical research; but he must have been delighted when Oscar Gross, in Germany, reported in 1914[25] that an enzyme capable of oxidizing homogentisic acid had been found in normal plasma, and that this enzyme was deficient in patients with alkaptonuria. Garrod incorporated these apparently confirmatory observations into the 1923 edition of *Inborn errors of metabolism*. In 1958, twenty-two years after Garrod's death, Bert N. La Du and colleagues, using more chemically sophisticated methods, showed that Gross's data were in error.[26] In 1962 it finally became clear that patients with alkaptonuria have a specific deficiency of a liver enzyme, homogentisic acid oxidase, which catalyzes the oxidation of homogentisic acid to maleyl acetoacetic acid.[27]

In 1907 Muriel Wheldale Onslow had already begun her studies at Cambridge University on the variation of color in *Antirrhinum* flowers, and its relationship to genes and enzymes.[28] She continued this work at the John Innes Horticultural Institution at Merton, near London, where Bateson became director in 1910. The chemistry of the anthocyanin pigments responsible for the coloration of *Antirrhinum* was extremely complicated, and although Onslow reviewed her work comprehensively in *The anthocyanin pigments of plants* in 1916, she did not refer to Garrod.[29] Nor, it should be said, did Garrod ever refer to her plant work.

In 1920 Haldane had speculated on the chemical nature of the gene, but did not refer to Garrod's work.[30] When Haldane left Cambridge in 1933 to take up a position at University College in London he had already established a working relationship with Onslow, and their relationship became more productive because of his part-time appointment at the John Innes. Taking great pride in having initiated the work on the anthocyanin pigments, he petulantly regretted having to share the credit for the work with Onslow:

Haldane maintained his interest and contribution of ideas to our work ... In later years his bone of contention was that his claim to have initiated and inspired the study of chemical genetics in flower colour had of necessity to be shared with Mrs. Onslow.[31]

In 1937, a year after Garrod's death, Rose Scott-Moncrieff, who also worked at the John Innes, summarized the existing knowledge of the relationship of genes to pigment production in flowers. In a phrase redolent of Garrod, she wrote, 'These genes which control series of chemical reactions are strong though novel weapons with which to attack the problems of biosynthesis.'[32]

When Onslow's book came out, Garrod was still stationed in Malta, and it is extremely unlikely that he would have known about her work. Bateson, however, with whom Garrod had kept in touch, was from the very beginning an enthusiastic supporter of the anthocyanin research. It might have been expected that Garrod, on his return home, would have caught up with Bateson and the plant work, which had provided such striking confirmatory evidence for Garrod's ideas. There is no evidence, however, that he did so.

In the United States, Thomas H. Morgan,[33] in company with the other early geneticists, paid little attention to the genetics of human disease. Although he had unwisely cited eye color as an example of a clear-cut Mendelian trait, he ignored, or more probably was unaware of, Garrod's *Inborn errors of metabolism*. When Morgan delivered his Nobel Lecture in 1934, he failed to mention Garrod. If he had read either *Inborn errors of metabolism* or *The inborn factors in disease*, published in 1931, it is doubtful that he would have written with such evident disdain:

I am aware, of course, of the ancient attempts to identify certain gross physical human types — the bilious, the lymphatic, the nervous, and the sanguine dispositions, and of more modern attempts to classify human beings into the cerebral, respiratory, digestive and muscular, or, more briefly, into asthenics and pycnics. Some of these are supposed to be more susceptible to certain ailments or diseases than are other types, which in turn have their own constitutional characteristics. These well-intended efforts are, however, so far in advance of our genetic information that the geneticist may be excused if he refuses to discuss them seriously.[34]

Sewall Wright, the population geneticist whose name is linked closely with Haldane and Ronald A. Fisher, had been a graduate student with William E. Castle at Harvard University, where he worked on the inheritance of coat color in guinea-pigs. Wright was probably the first American geneticist to appreciate the full significance of Garrod's work. In a letter to the author, Wright wrote:

I first became aware of Garrod's work on reading *Mendel's principles of heredity*, 1913 edition, probably in 1913, shortly after beginning graduate work with Prof. Castle. I probably did not read his own papers at that time, however, since I did not list him in the bibliography of my thesis published in 1916. In discussing "Hereditary factors and the physiology of pigment" in this, I find that I credited Cuénot with having made "the first attempt to correlate the facts of Mendelian inheritance with the physiology of the pigment".[35]

Between 1915 and 1925 Wright worked at the US Bureau of Animal Industry in Washington. He later recalled, with evident sadness, that, while there, he had no courses to teach and no one with whom he could discuss the physiological aspects of enzyme action. However, when he joined the Zoology Department of the University of Chicago, in 1929, Wright immediately introduced a course of 30 lectures, entitled 'physiological genetics'. Three of the lectures were devoted to Garrod and inborn errors of metabolism.[36]

With few exceptions, the major textbooks of biochemistry and genetics paid only token attention to Garrod's work on inborn errors of metabolism, and almost no one referred to his concept of biochemical individuality. Haldane and Julian S. Huxley,[37] in their book *Animal biology*, published in 1927, did not refer to Garrod. His name also remained absent from the 1937 edition of *The science of life*,[38] the immensely popular book written by H. G. Wells, Julian Huxley, and G. P. Wells that had originally appeared in print in 1929. Richard B. Goldschmidt,[39] a distinguished German geneticist, referred neither to Garrod nor to his work in his influential 1938 book, *Physiological genetics*.

The reasons for this neglect are numerous; but Lederberg suggests that the concept of protoplasm as a living, dynamic, colloidal system, and the prevailing opinion that cellular protoplasm was influential in determining gene action may have delayed the recognition that nuclear genes were primarily responsible for heredity.[40] This uncertainty had been reflected as late as 1934, in Morgan's Nobel Lecture 'The relation of genetics to physiology and medicine'.[41] Considering the lecture's title, it is, as Lederberg has again noted, quite remarkable that Morgan made no reference to Garrod's work. It becomes evident that leading zoologists and geneticists of the day were unwilling to commit themselves to any particular concept of how genes achieve their physiological effects.

It would be quite wrong to infer from this, however, that Garrod's work was entirely overlooked by physiologists. As early as 1898 Gowland Hopkins discussed, albeit without attribution, Garrod's work on alkaptonuria in a chapter he wrote on 'The chemistry of the urine' for Edward Schäfer's two-volume *Text-book of physiology*. Hopkins's description omitted any notice that the disease had a familial component, and he contented himself with the statement that:

Although thought to be especially frequent in various forms of tuberculosis, alkaptonuria must not be looked upon as specifically associated with any particular diseased conditions; it indicates rather some peculiar independent changes of metabolism, and is not infrequently met with in conditions of apparent health.[42]

Hopkins asserted, despite Garrod's growing evidence to the contrary, that the excess of homogentisic acid was probably due to the action of specific micro-organisms present in the bowel. This erroneous conclusion is the more surprising because Garrod and Hopkins were actively collaborating colleagues, in addition to being close friends.

In the first decade following the rediscovery of Mendelism, a number of medical textbooks, particularly in Germany, discussed the role of 'constitution' and, to a lesser extent, heredity in the causation of disease.[43]

In the 1907 edition of William Osler's classic work, *Modern medicine*, the first chapter was written by J. George Adami, Professor of Pathology at McGill University, who devoted 36 pages to a discussion of 'Inheritance and disease'.[44] In the section on nutrition and metabolism, which was written by Russell H. Chittenden, Professor of Physiological Chemistry at Yale, and his colleague Lafayette B. Mendel,[45] no reference was made to alkaptonuria or cystinuria. In neither article was there mention of Garrod.

In 1927 the American biochemist Meyer Bodansky, in the first edition of his textbook *Introduction to physiological chemistry*,[46] devoted several pages to alkaptonuria, and correctly concluded that 'homogentisic acid represents an intermediate in the normal metabolism of phenylalanine and tyrosine.' He also discussed cystinuria, and asserted that it, too, was a hereditary disorder. Although the first edition of Bodansky's book is well referenced, Garrod is not mentioned by name until the third edition, published in 1934.[47]

Although the concept of inborn errors of metabolism as freaks of nature had become relatively well recognized by the mid-1930s, Garrod's deeper and more important concept — that of chemical individuality — emerged only much later. Roger J. Williams had devoted a short paragraph to alkaptonuria in the 1931 edition of *An introduction to biochemistry*,[48] but he did not mention Garrod or discuss the biological implications of biochemical variation. In 1956, twenty years after Garrod's death, Williams wrote a highly insightful and influential text, *Biochemical individuality*,[49] in which he showed he clearly understood the importance of Garrod's concepts, and quoted from Garrod's 1902 paper on alkaptonuria and chemical individuality. Williams predicted, accurately, that with the introduction of paper chromatography and other methods of separating closely similar chemical compounds, the recognition of biochemical individuality would increase rapidly.

The first scientific acknowledgement that Garrod's concept of biochemical individuality was important may have come from Haldane in 1937, the year after Garrod's death, in a volume of 31 essays written by members of the Laboratory of Biochemistry at Cambridge University on the occasion of Gowland Hopkins's 75th birthday.[50] Haldane, who had worked with Hopkins for twelve years, chose the unambiguous topic 'The biochemistry of the individual', and clearly had Garrod in mind when he wrote:

The history of biochemistry shows that the study of individual abnormalities has been of enormous importance. Indeed, one whole branch of biochemistry arose from the study of human beings whose abnormalities had attracted the attention of men who thought along chemical lines.[51]

Later on, he refers to Garrod explicitly:

Genetics is concerned with the analysis of the innate differences between individuals. Garrod's pioneer work on congenital human metabolic abnormalities such as alcaptonuria and cystinuria had a very considerable influence both on biochemistry and genetics.[52]

The pre-eminence of Hopkins in the field of biochemistry guaranteed that this distinguished volume of essays, edited by Joseph Needham and David Green, would be widely read, certainly by biochemists.

In May 1935 George W. Beadle had gone to Paris to work in Boris Ephrussi's laboratory to investigate the enzymes that controlled the development of eye color in *Drosophila*. On his way back to California, in 1936, Beadle visited the John Innes Horticultural Institution, where Haldane had become Director and Scott-Moncrieff was continuing her work on the genetics and biochemistry of flower color. The main conversation, Beadle recalled, revolved around biochemistry, enzymes, and gene action, but, as he was later to write, 'Haldane said nothing about Garrod nor did any of the others'.[53]

By 1941 Tatum had introduced Garrod's work on comparative biochemistry into his course at Stanford University.[54] In 1942, five years after his paper for the *Festschrift* for Gowland Hopkins, Haldane seems still to have regarded genes as probably protein in nature ('The size of the gene is roughly that of a protein molecule'[55]); and as late as 1947, Herman Müller was still discussing the action of genes in terms of variations in protein structure.[56]

Garrod died six years before the publication of Haldane's *New paths in genetics*,[57] in which Haldane gave Garrod unambiguous credit for anticipating the one gene–one enzyme hypothesis. Beadle, in response to a query about when he had first learned about Garrod, wrote to the author that 'In my case it was in reading Haldane's *New paths in genetics* in 1942.'[58] However, Beadle had already referred to Garrod in a paper on *Drosophila* delivered to the American Association for the Advancement of Science in January 1941.[59] Moreover, in 1939 Beadle had mentioned Garrod in a paper given before the Seventh International Genetical Congress in Edinburgh.[60]

By 1946, when David Bonner[61] discussed the relationship between genes and enzymes in a paper on biochemical mutations in *Neurospora* at a meeting in Cold Spring Harbor, New York, the one gene–one enzyme hypothesis had not yet been entirely accepted. In opening the discussion, Max Delbrück was characteristically, but appropriately, skeptical:

The question arises whether the evidence obtained actually supports the thesis, beyond the mere fact that it is compatible with it. The possibility exists, and should be discussed, that the experimental approach applied in the Neurospora work is such that incompatibilities would be unlikely to arise even if the thesis were not true at all ... to make a fair appraisal of the present status of the thesis of a one-to-one correlation between genes and species of enzymes, it is necessary to begin with a discussion of methods by which the thesis could be *dis*proved. If such methods are not readily available, then the mass of "compatible" evidence carries no weight whatsoever.[62]

Lederberg also remained skeptical,[63] and in 1951 Norman Horowitz and Urs Leupold were still addressing Delbrück's concerns.[64] They had been able to show, using temperature-sensitive mutants of *Neurospora*, that in over 80 per cent of the mutations a single growth factor was lost. Thus, in contrast to *Drosophila*, lethality in *Neurospora* resulted from the loss of a single gene. In the same symposium, Bonner asserted that the one gene–one enzyme hypothesis not only had validity, but also represented a focal point for the convergence of genetics and biochemistry.[65]

The evolution of the one gene–one enzyme concept has been recently revisited by Horowitz, who also emphasizes that the hypothesis had the distinction of bringing the twin disciplines of biochemistry and genetics together in a fruitful way, and sparked the explosive development of biochemical genetics.[66]

More than half a century has now passed since Garrod died. Advances in our knowledge of genes and gene action have proceeded at dazzling speed, and no attempt will be made here to document them in any detail. The discovery by Oswald Avery, Colin MacLeod, and Maclyn McCarty in 1944 that genes are made not of protein, but of DNA, set the stage for James D. Watson and Francis H. C. Crick to establish the structure of DNA and thus to provide a mechanism for replication. From these basic observations the science of genetics and its applications to man have grown into one of the most vigorous contemporary areas of biological research.[67]

Several thousand inborn errors of metabolism have been identified, and the number will continue to grow rapidly. The implications of this knowledge for the prevention and treatment of inherited disease are becoming increasingly evident. While the one gene–one enzyme hypothesis has been expanded from its original definition, and the mechanism for single-gene defects is largely understood, the polygenic basis of many common diseases, including infectious diseases, hypertension, and cancer, is only now being explored at the molecular level. Disentangling the molecular basis of how genes interact, both with each other and with the environment, will be a spectacular chapter in the evolution of our knowledge of the role of genes in the causation of many common diseases.

Garrod's last contribution, *The inborn factors in disease*, provided the intellectual basis for his belief that the role of genes must be considered in all human disease, and for his clear recognition that biochemical variation was of profound evolutionary significance.

Today, few doubt Garrod's contributions. At a time when internal medicine was beginning to fragment into subspecialties (he himself first followed his father in specializing in rheumatic diseases and gout),

Garrod continued to see medicine as a whole. He was able to put biological variability and disease into an evolutionary perspective, and it is only surprising, as Childs has long emphasized,[68] that this perspective has not yet influenced medical education. Even now, although human genetics is represented in most medical schools, departments are rare, and most medical educators believe that human genetics need not play a major role in the curriculum. Garrod saw it otherwise. He believed that only in the context of biochemical individuality would human disease be understood.

13

A Summing Up

Those who know the end of the story can never know what it was like
at the beginning.

— C. V. Wedgwood

GEORGE Beadle's 1958 Nobel Lecture firmly ensconced Garrod's ideas
in the history of scientific medicine as the precursor of the one
gene–one enzyme hypothesis.

Beadle had very particular motives for reaching back to Garrod's
work. When he and Edward L. Tatum articulated the hypothesis that a
single gene only governed the synthesis of a single enzymatic protein,
they met with some legitimate skepticism, particularly from Max
Delbrück,[1] the acknowledged doyen of the field, and Joshua
Lederberg.[2] Beadle and Tatum buttressed their position by pointing out
that their 'new' hypothesis was in fact rather old and well established.
Garrod's discovery that alkaptonuria was the manifestation of a
recessively inherited disease characterized by the absence of an enzyme
was too good not to use.

By 1958, Garrod's work on alkaptonuria and other inborn errors of
metabolism had assumed a permanent though inconspicuous place in
most medical and biochemical textbooks, so there is no great mystery
about how Beadle and Tatum could have known about it. Although
they quoted Garrod to advantage, they did not explore the full force of
his medical and biological insights.

It was not that the fruits of Garrod's research did not support Beadle
and Tatum's one gene–one enzyme hypothesis. They did. Moreover,
Beadle and Tatum's hypothesis was specifically and narrowly concerned
with the one-to-one relationship between genes and enzymes. On the
other hand, Garrod's central insight was not simply the absence of
enzymes in recessively inherited disease. For him, these rare diseases
were merely one set of examples that demonstrated the fundamental
principle of biochemical individuality. It was this broad concept that I

believe Beadle and Tatum may have under-appreciated. Biochemical individuality was not, after all, central to their argument; their principal commitment was to a direct experimental approach to research. Beadle and Tatum were not physicians, and had not had the experience of witnessing, first-hand, the complex and variable expression of the same human disease in different individuals. Because man was such an unpromising experimental organism, they had relatively little interest in the latent field of human genetics. It would be equally unwarranted to presume that Garrod himself, although he had recognized a relationship between genes and enzymes, had displayed any discerning insights into the singularity of Beadle and Tatum's emerging one gene–one enzyme hypothesis.

It has been said that after the First World War and the tragic loss of his three sons, the fire went out of Garrod. To add to this unrelenting personal sorrow, he may well have become increasingly dispirited by his utter inability to influence the thinking of his medical contemporaries, despite their unconcealed admiration for his scholarly approach to medicine and human disease. When they referred to his work, the acknowledgement was usually minimalistic, a token genuflection to the author of *Inborn errors of metabolism*, identifier of rare metabolic diseases that most of them would never encounter in their professional lives.

When Garrod went to Oxford, he left behind the bustling and demanding world of consulting practice. He did so with some relief, for the daily grind of seeing patients was time-consuming and exhausting, and left him little opportunity to develop his ideas on chemical individuality. As Regius Professor of Medicine, he would have no need to earn his living by consulting practice, and his clinical duties at the Radcliffe Infirmary would be light. Garrod enjoyed teaching, and the administrative duties he was asked to assume, though consequential, were not over-burdensome. He hoped in the relative tranquility of Oxford academic life to marshal his ideas in what he knew would be his last chance to get his message across. With the publication of *The inborn factors in disease* in 1931, Garrod produced a prescient medical masterpiece. Ironically, this book, which contains some of Garrod's clearest, most graceful writing, was almost totally ignored.[3]

For nearly forty years, Garrod had enjoyed a warm personal and professional relationship with Gowland Hopkins. They had worked together in London on urinary pigments, and it was only in 1898, when Michael Foster recruited Hopkins to Cambridge to lead the Department of Biochemistry, that their professional collaboration ended. Hopkins was recognized as the foremost biochemist of his day, and would become a Nobel Laureate. He and his colleagues in the Department of Biochemistry, particularly J. B. S. Haldane, had a long-time interest in the nature of enzymes and their mode of action. Garrod's retirement to Cambridge provided a ready opportunity for him to expound to his friend his ideas on biochemical individuality. Yet even though Garrod must have talked to Hopkins on numerous occasions, there is little evidence that Hopkins or other biochemists understood the biological significance of what Garrod was saying. Hopkins wrote a sympathetic and generous obituary of Garrod for the Royal Society, but even there he made no reference either to *The inborn factors in disease* or to Garrod's concept of biochemical individuality. Perhaps Hopkins, whose genius was rooted in experimental work, believed that the whole concept of biochemical individuality was so much metaphysical talk.

Hopkins pursued the experimentally possible, marching steadily onward from one brilliant achievement to the next. His scientific habits were established, secure, and immensely successful. Garrod's soliloquies led nowhere experimentally. He could not suggest experiments for Hopkins to undertake; he merely offered his conviction, expressed over and over again, that the clue to evolution, to rare inborn errors of metabolism, and even to common human diseases would be found in a detailed understanding of biochemical individuality.

In the 1950s, this concept was still scarcely recognized. Few biochemists realized that the structure of bovine ribonuclease, so brilliantly elucidated by Christophe H. W. Hirs, Stanford Moore, and William H. Stein,[4] necessarily represented the *mean* structure of the enzyme. The starting material for their structural studies on ribonuclease came from material that had been obtained by extracting the enzyme from the pancreases of many animals. If, for example, 5 to 10 per cent of these animals possessed a genetic variant of pancreatic ribonuclease that differed from the 'normal' by a single amino acid, that difference would not have been detected. It is only in the last twenty-five years that biochemical polymorphisms, with their attendant biological and evolutionary significance, have been recognized.

Although Garrod was not particularly gregarious, he enjoyed discussing his work with colleagues. But when he did so, he was evoking another world. He was asking his colleagues to extrapolate ideas from an earlier, Darwinian epoch and to reflect on their evolutionary consequences in biochemical terms. To do this, it was necessary for his listeners to put on a new thinking-cap, to use Herbert Butterfield's phrase.[5] But a new thinking-cap is not enough, for that plays down the practical life of science. It takes more than a good idea — it takes knowing what to do next. This connective emphasis is displayed in Ian Hacking's *Representing and intervening*, in which the necessity of moving from the armchair to the laboratory bench is made explicit.[6] Garrod never doubted the existence of biochemical individuality, but he could not, to use the evocative phrase made popular by Hacking, 'spray it with red paint'.[7]

In discussing biochemical individuality, Garrod was saying, in essence, 'don't map diseases to the organs they involve, map diseases to the individuals they afflict'. During his life, he had assembled plenty of evidence that this would be a profitable way to approach disease and susceptibility to disease. Garrod's concept of biochemical individuality was a purely personal scientific revolution; Hopkins did not have the tools with which to confirm or reject the general applicability of the hypothesis. There was simply nothing he could do to corroborate his friend's theory. With the arrival of increasingly effective separatory technologies, such as chromatography and electrophoresis, some thirty years later, it became possible to separate closely similar proteins in ever-miniaturizing detail.

* * *

It would be quite impossible to summarize here the advances in genetics, particularly human genetics, that have occurred since Garrod's death in 1936.[8] The recognition that DNA was the genetic material by Oswald Avery, Colin MacLeod, and Maclyn McCarty in 1944, and the establishment of the structure of DNA, which provided a mechanism for its replication, by James Watson and Francis Crick in 1953 are the soaring peaks in a mountain range of astonishing discoveries. This summing up is restricted to some of the developments in the two fields that interested Garrod most — inborn errors of metabolism and biochemical individuality.

In his 1908 Croonian Lectures Garrod had discussed, as examples of inborn errors of metabolism, albinism, alkaptonuria, cystinuria, and pentosuria. Since his death, a number of significant advances have been made in our understanding of these disorders. In his chapter on albinism, Garrod concluded:

... the lines along which the systematic study of albinism may profitably be directed are only beginning to indicate themselves. The carrying out of such a research remains as a task for future workers.

There is now increasing evidence that albinism, along with most inborn errors of metabolism, is genetically heterogeneous. Two major forms of the disease are now recognized: ocular albinism, the one described by

Garrod (inherited as an autosomal recessive), and the X-linked oculocutaneous form of the disease, now thought to be the most common variety. Garrod and Bateson were both aware of the latter, which was first described by the ophthalmic surgeon Edward Nettleship in 1908. Additional variants of albinism are now recognized, including a second non-allelic recessive form of the disease. In all, some ten variants are now defined by their clinical and biochemical characteristics, as well as their tyrosinase activity. Our knowledge of alkaptonuria since Garrod's death has advanced surprisingly slowly. The occasional pedigree suggesting that the trait could be inherited as a Mendelian dominant as well as a recessive was interpreted by Garrod as being due to the occasional marriage of a homozygous with a heterozygous individual. This view is supported by the extensive studies of Milch[9] and alkaptonuria remains the classic example of a recessively inherited inborn error of metabolism.

Cystinuria has been found to be extremely complicated since Garrod's death, and I will not attempt to provide an up-to-date account of the disease. Garrod's view that it could be regarded as a disturbance in cystine metabolism due to the absence of an enzyme must be modified. Today, cystinuria is usually regarded as a primary disturbance in the cell membrane concerned, which affects the renal and intestinal transport of cystine, lysine, and arginine. The three known variants of cystinuria are considered allelic mutations, some of which are expressed in the heterozygous state. The possibility that disordered cystine and dibasic amino acid transport may ultimately be explained by an enzymatic deficiency affecting transport cannot be excluded.

Garrod's observations on pentosuria, the fourth of his inborn errors of metabolism, have been amply corroborated. The condition is, as he suggested, recessively inherited, is almost exclusively confined to Ashkenazi Jews, and has no discernible functional consequences. The metabolic error is now known to be caused by the absence of the enzyme L-xylose reductase, which catalyzes the conversion of L-xylose to xylitol.

When Garrod wrote the second edition of *Inborn errors of metabolism*, in 1923, he included two additional conditions, hematoporphyria congenita (congenital porphyrinuria) and congenital steatorrhea. Disorders of porphyrin metabolism have been the subject of intense research since Garrod's death. Although his description of hematoporphyrinuria congenita requires little in the way of modification, the complexity of the condition has become increasingly apparent. As Garrod recognized, an attack of porphyria can be precipitated by light, by taking the hypnotic drug sulfonal, and by the ingestion of other drugs. The specific enzymatic deficiencies responsible for at least eight varieties of porphyria have been identified. Some of these disorders appear to be inherited in a dominant fashion, whereas others, such as congenital erythropoietic porphyria, which Garrod had termed hematoporphyria congenita, are inherited, as he had shown, as autosomal recessives.

The decision by Garrod to include congenital steatorrhea as an inborn error of metabolism was less secure. On the basis of three patients, two of whom came from a family in which the parents were first cousins, Garrod postulated that the steatorrhea in the family was probably a recessively inherited disorder of fat metabolism. Although this possibility cannot be excluded, subsequent studies have failed to uncover similar cases. Steatorrhea, however, is not uncommon in childhood, and familial features have been noted. It has recently been suggested that part of the genetic predisposition to idiopathic steatorrhea may be determined by alleles at the HLA-B8 or DW3 locus and an antigen on B lymphocytes.

The first unequivocally demonstrable enzymatic deficiency in a human disease — an inherited deficiency of the NADH-dependent enzyme responsible for the reduction of methemoglobin, leading to methemoglobinemia — was described by Alexander G. Gibson in 1948, twelve years after Garrod's death. Garrod had, of course, postulated that many inborn errors of metabolism would be found to be caused by a deficiency of a protein with a specific enzymatic function; but it has now become abundantly clear that a number of inherited molecular defects may occur in any protein.

In 1949 Linus Pauling, on the basis of the electrophoretic differences between normal and sickle-cell hemoglobin, suggested a new category of disease — molecular disease; and seven years later Vernon A. Ingram identified the molecular difference between sickle-cell hemoglobin and normal hemoglobin.[10] Garrod would be startled to learn that today, with the application of sophisticated techniques of nucleic acid chemistry, about which he would have known nothing, the mutations that give rise to the sickle-cell phenotype can be identified at the nucleotide level. He would marvel at the knowledge that the disease can be identified, prenatally, by examining as little as 15 milliliters of uncultured amniotic cells, and that more than 170 sites have now been identified in the betaglobin chain that, when mutated, can give rise to specific disease phenotypes. As a physician, however, Garrod would doubtless be chagrined to learn that there is still no effective therapy for a disease that has been studied at the molecular level for more than thirty years.

Even though Garrod predicted that the four inborn errors of metabolism he identified and described in his Croonian Lectures were only the first of many such disorders, the sheer number of the inborn errors of metabolism that have been uncovered during the last twenty-five years is remarkable.

Keeping track of the almost weekly recognition of new inborn errors of metabolism and genetic traits is a formidable task, and human geneticists are greatly indebted to Victor A. McKusick, who began in 1960 to catalog all the genetic information on the human X chromosome. From this modest beginning, the McKusick catalog of X-linked, dominant, and recessive phenotypes has grown to 5710 entries.[11] These entries include large numbers of polymorphic variations, only some of which are associated with disease.

Garrod would have welcomed the publication in 1960 of *The metabolic basis of inherited disease*, in which all the major inborn errors of metabolism then known were discussed. Now, in its sixth edition, this classic and massive work can discuss, in addition to diseases caused by single genes, conditions that are influenced by the environment and the expression of other genes.[12] In the next edition, such conditions will very probably receive additional emphasis. For Garrod, perhaps the best of good news would have been the publication, in 1992, of a new work entirely devoted to the genetics of common diseases.[13]

Although Garrod preached the importance of human biochemical individuality with the stamina, if not the fire, of John Knox, he had little in the way of biochemical evidence to support his homily. The experimental evidence for his visionary concept is now overwhelming, and protein polymorphisms, which have been extensively studied by Harry Harris and his co-workers [14] in man, and later in *Drosophila* by Richard C. Lewontin,[15] are being increasingly investigated at the level of DNA.

Garrod's approach to the elucidation of inherited metabolic disease began with the recognition of a phenotypic difference. Now, in addition to this classic approach, it is possible, using DNA technology, to move from the gene to the phenotype — the so-called reverse, or positional, genetics that proved so successful in identifying the defect in DNA responsible for Duchenne muscular dystrophy and cystic fibrosis.

Garrod would probably be disappointed to learn how little concrete biochemical information concerning man's susceptibility to specific infectious agents has been uncovered. Tuberculosis claimed several members of his family, including three of his siblings, and it would have fascinated him to learn that a single gene on chromosome 1 in the mouse determined susceptibility to tuberculosis. Garrod, in addition to pondering its human homologue, would have wanted to know the precise nature of the biochemical and immunological basis underlying the susceptibility.[16]

The molecular complexity of human leukocyte antigens (HLAs), their role in the human antigenic response, and the part they play in determining susceptibility to disease would have attracted his attention. Early in his career, he established a reputation in the field of rheumatic disease, and the finding, for instance, that 90 per cent of patients with ankylosing spondylitis possess a specific leucocyte antigen, HLA-B27, would have provided him with a molecular update on the relationship between genetic constitution and this disease. However, despite the massive research now devoted to the HLA locus and its relationship to a variety of diseases, the exact molecular mechanism underlying the association of a particular HLA type and disease remains elusive.

The international project for mapping the human genome would have amazed Garrod. To date, 6000 of the approximately 100 000 human genes have been identified, leaving over 90 per cent of the human genome still uncharted. Most genes consist of between 10 000 and 150 000 nucleotides, not all of which code for proteins; none the less, obtaining the complete sequence of human DNA will require identifying approximately 3000 million nucleotides. With the development of increasingly rapid and sophisticated sequencing methodology it is possible that within ten to fifteen years, or sooner, as some believe, we will have a complete road-map of the human genome — a DNA dictionary.[17]

Garrod would have welcomed this information, not only for the identification of disease-causing genes, but also for the identification of genes that lead to susceptibility to a variety of common conditions, such as cancer, behavioral disorders, and hypertension. However, he would, I believe, still regard the information, vast though it will be, as a mere beginning, a basis for posing new and interesting biological questions.

The nature and extent of human genetic diversity, Garrod's intellectual holy grail, will not be known even when the human genome is sequenced. I have a hunch, however, that when the human genome is sequenced, we will not only begin to address problems of human disease and evolution from an entirely new perspective, but we will begin to understand the functions of those genes that do not code for protein. This DNA, sometimes irreverently called 'junk DNA', will, I have no doubt, prove to have marvelous and unanticipated biological functions.

Perhaps when the year 2050 comes along, almost two hundred years after the birth of Archibald Edward Garrod, we will have begun to understand, in exquisite molecular detail, the polymorphic variation in the nucleotide sequences of man, and to comprehend the extent and importance of human biochemical diversity in all its functional and evolutionary significance. Yet new horizons will continue to emerge from the clouds of ignorance, for the search for deeper understanding of man and his diseases will continue until time ceases. Garrod would approve.

NOTES AND REFERENCES

Chapter 1

1. For an interesting account of the appalling sanitation and health that existed in Victorian London, see Anthony S. Wohl (1983). *Endangered life: public health in Victorian Britain.* Harvard University Press, Cambridge, Mass.; and F. B. Smith (1990). *The people's health, 1830–1910*, Weidenfeld & Nicolson, London. A useful overview of the political and social conditions in England between 1815 and 1895 can be found in Derek Beales (1969). *From Castlereagh to Gladstone 1815–1855*, Norton, New York. For those who wish to delve more deeply, R. C. K. Ensor (1987). *England 1870–1914.* Oxford University Press, Oxford, is highly recommended, as is Sir Llewellyn Woodward (1962). *The age of reform 1815–1870*, (2nd edn). Oxford University Press, Oxford.
2. Van Leeuwenhoek, A. (1683). Letter to the Royal Society, quoted in R. Dubos, (1962). *The unseen world*, Rockefeller Institute Press in association with Oxford University Press, New York.
3. Peterson, M. J. (1978). *The medical profession in mid-Victorian London*, p. 179. University of California Press, Berkeley; Peterson, M. J. (1984). Gentlemen and medical men: the problem of recruitment, *Bulletin of the history of medicine*, **58**, 457–73. For a scholarly account of the evolution of hospital care during the professional careers of Archibald Garrod and his father, Alfred Baring Garrod, the treatise of Brian Abel-Smith (1964), *The hospitals, 1800–1948*, Heinemann, London is highly recommended. For a detailed discussion of the life and backgrounds of physicians in Victorian England, Peterson, M. J. (1978), *The medical profession in mid-Victorian London*, University of California Press, Berkeley, is essential reading. The nursing conditions that prevailed are well documented by Summers, Anne (1989). The mysterious demise of Sarah Gamp: the domiciliary nurse and her detractors, *c.* 1830–1860. *Victorian Studies*, **32**, 365–86.
4. Alfred Baring Garrod (1819–1907) was the only son of Robert Garrod and his wife, Sarah Ennew Clamp. Alfred Garrod is sometimes known as the father of rheumatology, and was one of the first physicians in London to raise his consultation fees from one to two guineas. His interest in materia medica and therapeutics remained strong even after he had retired from practice. He was editor of the British *Pharmacopoeia* for many years. In Aix-les-Bains they named a street 'Rue Sir Alfred Garrod' in his honor. When he died, he left his son Archibald £19 500. Obituaries were published in the *British Medical*

Journal (1908), **121**, 58–9, and the *Journal of the American Medical Association* (1973), **224**, 663–5.

5. An obituary of Robert Garrod was published in the *East Anglian Daily Times,* 12 June 1877.

6. Garrod, A. B. (1848). Observations on certain pathological conditions of the blood and urine in gout, rheumatism and Bright's disease. *Medical and Chirurgical Transactions,* **31**, 83–97.

7. Garrod, A. B. (1855). *The essentials of materia medica, therapeutics, and the pharmacopoeia.* Walton & Maberly, London.

8. Garrod, A. B. (1859). *The nature and treatment of gout and rheumatic gout.* Walton & Maberly, London.

9. Garrod, A. E. (1890). *A treatise on rheumatism and rheumatoid arthritis.* Griffin, London. In 1891, this was translated into French by Léon Brachet as *Traité du rhumatisme et de l'arthrite rhumatoïde,* Paris.

10. Alfred Henry Garrod (1846–79), the eldest child of Alfred Baring Garrod, was born at 9 Charterhouse Square, where his father practised before moving to Harley Street. A collection of his scientific papers was published posthumously: Garrod, Alfred Henry (1881). *In memoriam. The collected scientific papers of the late Alfred Henry Garrod* (ed. W. A. Forbes). R. H. Porter, London. A useful memoir can be found in the *Dictionary of national biography* (1968), p. 906, Oxford University Press, London.

11. The name tripos is given to these examinations because originally candidates sat on three-legged stools.

12. Herbert Baring Garrod (1849–1912) was called to the bar after graduating from Oxford, but his interests remained in literature rather than the law. In 1884, he married Lucy Florence Colchester, daughter of William Colchester of Springfield, Ipswich. In 1886, he was appointed General Secretary of the Teachers Guild of Great Britain and Ireland, and devoted his life to its work. He also formed the United Profession of Teachers, a controlling body similar to the General Council of the Bar and the General Medical Council.

13. Garrod, O. (30 November 1979). Letter to A. G. Bearn. Private and unpublished.

14. Darwin, F. (ed.) (1898). *The life and letters of Charles Darwin,* Vol. I, p. 453. Appleton, New York.

15. Fruton, J. S. (1972). *Molecules and life, historical essays on the interplay of chemistry and biology.* Wiley-Interscience, New York.

16. Until the end of the nineteenth century, the most usual treatment for scarlet fever was 'antimonial wine' in the early stages, followed by 'cupping and counter irritants on the skin' toward the end. The ulceration of the throat that accompanied the more virulent forms of the disease could be checked by 'general evacuants and copious blood-letting'.

17. Archibald Campbell Tait 1910–1991, *Encyclopaedia Britannica,* (11th edn), p. 363. Graham, George (1962) [a junior colleague of Archibald Garrod at St Bartholomew's Hospital]. Letter to F. Vella (7 October 1962). Private and unpublished.

18. Charles Samuel Keene (1823–91) was born in London but lived in Ipswich and attended Ipswich school, as did Archibald's father, Alfred Garrod. Keene's

mother, Mary, was the daughter of John Sparrow, of the so-called Ancient House in the Buttermarket, Ipswich, parts of which go back to the fifteenth century. It was through his mother that Keene was related to the Garrods. A charming and eccentric bachelor, he always carried an inkpot in his waistcoat and a sketchbook in his hand, and he played the bagpipes, alone and at night, on the Suffolk coast. Despite his eccentricities, he was a skilled draughtsman whom Whistler regarded as the greatest English artist since Hogarth. He was admired by Degas and Pissaro, and he was a friend of the poet Edward Fitzgerald. Archie's childhood interest in sketching may have been inspired by Keene's humorous drawings (see Layard, G. S. (1892). *The life and letters of Charles Samuel Keene*. S. Low, Marston, London). An exhibition of Keene's work was mounted by Christie's of London in January 1991. An entertaining review was published by Russell Davies (1991). *Times Literary Supplement*, 1 February, p. 14.

19. Meredith White Townsend (1831–1911) was born on the border between Essex and Suffolk and was educated at Ipswich Grammar School, as were Archibald's father and Charles Keene. Between 1864 and 1911 he lived in Harley Street, and was thus a close neighbor of Alfred Garrod. Townsend's wife, Alicia, was a relative as well, being the daughter of John Sparrow. *The Spectator*, of which her husband was co-editor, supported the North in the US Civil War, which angered the well-to-do in England, who favored the South.
20. Galton, F. (1869). *Hereditary genius*. Macmillan, London.
21. Garrod, A. E. (n.d., *circa* 1858). *The handbook of clasical* (sic) *architecture*. Private and unpublished.

Chapter 2

1. Garrod, A. E. (20 May 1868). Letter to his mother. Private and unpublished.
2. Garrod, A. E. (26 May 1868). Letter to his mother. Private and unpublished.
3. Garrod, A. E. (1869). Letter to his mother. Private and unpublished. This apparent disparity in the ratio of males to females has recently been explained by Sir Cyril Clarke: 'The sex ratio in butterflies is usually 1:1 but the females are much less in evidence and therefore collectors usually think there is an upset in the ratio.' Clarke, C. Letter to A. G. Bearn (19 August 1975). Private and unpublished. Plainly, the same observations that puzzled Archibald at the age of twelve trouble the amateur collector today. Quoted in Bearn, A. G. (1975–6). Lettsomian Lectures I. Archibald Garrod and the birth of an idea. *Medical Society Transactions*, **92**, 47–56.
4. Garrod, A. E. (1869). *The tiger*. Private and unpublished.
5. Garrod, A. H. Letter (14 November 1868) to Archibald E. Garrod. Private and unpublished.
6. Garrod, A. E. (1918). A tardy tribute. *The Marlburian*, Vol. **53**, No. 791, p. 164.
7. A discussion of the British public schools, particularly their traditions and customs, can be found in F. A. M. Webster (1937). *Our great public schools*, Ward Lock, London; and A. C. Benson (1906). *From a college window*, Putnam, New York. For details on Marlborough, one cannot do better than

consult A. G. Bradley, A. C. Champneys, and J. W. Baines (1923). *A history of Marlborough College*, John Murray, London.

8. Anon. (initials S.P.) (1904). Gilmore memorial notice, *The Marlburian*, Vol. **39**, pp. 71–2.

9. Benson, A. C. (1906) *From a college window*. Putnam, New York.

10. Marlborough School report card (summer term 1874). Private and unpublished.

11. Garrod's dismal record at school, coupled with his inability to achieve any success in the classics, resembles that of William Bateson, who was to become one of the most distinguished biologists of his time, and who drew Garrod's attention to the relevance of the work of Mendel. In each instance, teachers expressed dismay and unambiguous disapproval of their student's work. Quoted in Bateson, W. (1928). *William Bateson, F.R.S.: his essays*, p. 8. Cambridge University Press. Bateson's headmaster at Rugby, T. W. Jex-Blake, wrote in his report:

   Your son's work is far inferior to its true level: and for him to be 25th in Classics is scandalous. His Divinity with me is very poor; and in short the whole result is poor. Unless the two next Terms are wholly different I cannot advise you to send him back after the summer holidays; and it is very doubtful whether so vague and aimless a boy will profit by University life.

12. Garrod, A. E. (1918). A 'Tardy tribute'. *The Marlburian*, Vol. **53**, No. 791, p. 164.

13. Farrar, R. (1904). *The life of Frederic William Farrar, D.D., F.R.S., Sometime Dean of Canterbury*. Crowell, New York.

14. Marlborough School report card (Lent term 1875). Private and unpublished.

15. E. F. F. Chladni (1756–1827) was a German physicist who devised a method of analysing sound by spreading sand over plates of glass and running a violin bow over their edge — a popular demonstration in acoustics for schoolboys of the time.

16. Garrod, A. E. (1873–5), *The Marlburian*. In the records of the Natural History Society, Marlborough College.

17. Geissler's tubes were glass tubes with electrodes melted into the ends and filled with rarefied gases; they were employed to demonstrate induced currents.

18. R. W. Bunsen (1811–1899) and G. R. Kirchhoff (1824–1887). *Dictionary of scientific biography* (1981). Scribners, New York.

19. West, D. R. C. Letter to A. G. Bearn (6 February 1989). Private and unpublished.

20. *Proceedings of the Royal Geographical Society*, (1874–5), **19**, 526–36. 'The examination and the prizes began in 1868 with the intention of promoting the study of geography in schools but in 1883 they were abandoned as it was found that almost all the winning competitors came from Dulwich College and Liverpool College.' Kelley, C. Letter to A. G. Bearn (1 August 1989). Private and unpublished. Garrod's answers to questions have (remarkably) been preserved by the Royal Geographical Society.

21. Garrod, A. E. (1929). The power of personality. *British Medical Journal*, **2**, 509–12.

Chapter 3

1. The term 'Commoner' is applied to undergraduates at Oxford whose entering academic performance is not of a sufficiently high standard for them to be awarded scholarship aid from their college. They now form the majority of the undergraduate body.

2. Garrod, A. E. (27 April 1880). Letter to his mother. Private and unpublished.

3. Garrod, A. E. (25 October 1880). Letter to his parents. Private and unpublished.

4. A. G. Vernon Harcourt (1834–1919) was the nephew of one of the founders of the British Association. Harcourt was trained as a classicist, but on entering Balliol College he became a chemist, and furthermore contributed to contemporary understanding of the law of mass action. Although he was elected President of the Chemical Section of the British Association in 1875, his reputation at Oxford rested as much on his teaching skills as on his science. Obituary (1920): *Proceedings of the Royal Society*, **97**, 7–11.

5. Garrod, A. E. (1882). *The nebulae: a fragment of astronomical history.* Parker, Oxford (Johnson Memorial Prize essay of 1879; reconstructed and newly written, 1881). Herbert Friedman, astrophysicist, writes, 'Garrod's paper represented an unusually scholarly paper for an undergraduate ...'. Letter to A. G. Bearn (1 February 1991). Private and unpublished.

6. Medvei, V. C. and Thornton, J. L. (ed.) (1974). *The Royal Hospital of Saint Bartholomew.* Cowell, Ipswich.

7. Newman, C. (1957). *The evolution of medical education in the nineteenth century.* Oxford University Press, London.

8. Medvei, V. C. and Thornton, J. L. *Royal Hospital.*

9. Garrod, A. E. (1930). St. Bartholomew's fifty years ago: summer sessional addressed to the Abernethian Society, 5 June 1930. *St Bartholomew's Hospital Journal*, **37**, 179–82.

10. When Garrod graduated in 1884, wealthy patients, or those who were prepared to pay for medical care, were treated at home. Minor surgery was undertaken in the consulting rooms of the surgeon, who admitted his patients to a nursing home when extensive surgery was contemplated. However, from about 1875 a system of hospitals, known as Home Hospitals, was established for patients who were willing to pay for their treatment.

11. Allbutt, T. C. (1867). Letter: A Clinical Thermometer. *Med. Times & Gazette*, **i**, 182–3. T. Clifford Allbutt (1836–1925). Distinguished medical practitioner in Leeds; later, Regius Professor of Physic, Cambridge (1892–1925).

12. Garrod, O. (1938). The life of Samuel Jones Gee, M.D., F.R.C.P. (1839–1911). *St Bartholomew's Hospital Reports*, Vol. 71, pp. 229–79.

13. Garrod, A. E., *Fifty years ago*, pp. 200–4.

14. Garrod, O., *Life of Gee.*

15. Harvey, W. (1657). Annual list of the fellows and members of the Royal College of Physicians of London, p. 241. Words used by William Harvey in the conveyance of a gift to the Royal College of Physicians, 21 June 1656.

16. Gee, S. J. (1899). The Lumleian Lectures. Lecture I. Bronchitis, pulmonary emphysema, and asthma. *Lancet*, **i**, 743–7.

17. Garrod, O., *Life of Gee.*

18. Swirling, bone-chilling fog wreaked havoc with the elderly and debilitated, and it also provided a sinister and mysterious background for many of the fictional exploits of Sherlock Holmes, who at one time was reported by Conan Doyle to be working in the chemical laboratories at Bart's. Anderson, A. B. (1958). Teachers of chemistry and chemical pathology at Saint Bartholomew's Hospital. *St Bartholomew's Hospital Journal,* **62**, 311–14.

19. Jones, H. L. (1895). *Medical electricity: a practical handbook for students and practitioners.* H. K. Lewis, London.

20. Garrod, A. E. (1884). A visit to the leper hospital at Bergen (Norway), abstract. *St Bartholomew's Hospital Reports,* **30**, 311–13.

21. Although leprosy was well recognized by Avicenna, and was known to have occurred in Norway since antiquity, the disease experienced a resurgence in the eighteenth and nineteenth centuries. In 1855 there were 3000 recorded cases of leprosy in Norway, significantly fewer than the 6000 Garrod claimed. He also claimed that the disease would become extinct in 50 years (1934). His prediction, although wrong, was in the right direction: there were only 11 cases of leprosy known in Norway in 1950. Although the fish theory was favored by many, it is interesting that Boeck did not subscribe to it: Moore, S. W., quoted in Tebb, W. (1893). *Recrudescence of leprosy and its causation.* Sonnenschein, London.

22. MacDonnell, H. (1889). Note on leprosy in Norway and their special hospitals, *Lancet,* **ii**, 425.

23. Garrod, A. E., *Visit to the leper hospital.*

Chapter 4

1. Garrod, A. E. (1886). *An introduction to the use of the laryngoscope.* Longmans Green, London. Max Neuburger (1943) has given an excellent account of the importance of the Vienna school in his book *British medicine and the Vienna school: contacts and parallels,* Heinemann Medical Books, London; E. H. Majer and M. Skopec (1985) discuss the subject in their book *History of oto-rhino-laryngology in Austria,* Verlag Christian Brandstätter, Vienna.

2. *Munk's Roll,* Lives of the Fellows of the Royal College of Physicians of London 1826–1925, **4**, 179.

3. Duckworth, D. (31 March 1886). Letter to A. E. Garrod. Private and unpublished.

4. Garrod, A. E., and Garrod, L. (1866). Various letters to his parents. Private and unpublished.

5. Ibid.

6. Ibid.

7. Bridges, R., quoted in A. E. Garrod. (1930). St Bartholomew's fifty years ago. *St Bartholomew's Hospital Journal,* **37**, 200.

8. Herringham, W. P., Garrod, A. E., and Gow, W. J. (1894). *A handbook of medical pathology: for the use of medical students at St Bartholomew's Hospital.* Baillière, Tindall & Cox, London.

9. Naunyn, B. (1896). *A treatise on cholelithiasis.* (trans. A. E. Garrod). New Sydenham Society, London.

10. Garrod, A. E. (1897). Über den Nachweis des Hämatoporphyrins im Harn. *Zentralblatt Inneren Medizin*, **18**, 497–9.

11. Breda, A. (1897). A contribution to the clinical and bacteriological study of the Brazilian framboesia or "boubas" by Achilles Breda. In *New Sydenham Society: selected essays and monographs*, (trans. A. E. Garrod), pp. 259–83. New Sydenham Society, London.

12. Charlouis, M. (1897). On papilloma tropicum (framboesia). In *New Sydenham Society: selected essays and monographs*, (trans. A. E. Garrod), pp. 285–319. New Sydenham Society, London.

13. Garrod, A. E. (1890). *A treatise on rheumatism and rheumatoid arthritis*. Griffin, London. Translated into French (1891) as *Traité du rhumatisme et de l'arthrite rhumatoïde*, by Dr. Léon Brachet, Paris.

14. Garrod, A. E., and Cooke, E. H. (1888). An attempt to determine the frequency of rheumatic family histories amongst non-rheumatic patients. *Lancet*, **ii**, 110.

15. Ibid.

16. Higgins, T. T. (1952). *Great Ormond Street (1852–1952)*. Oldhams Press, London.

17. Letter from Robert Hutchison to C. J. R. Hart. Quoted in Hart, C. J. R. (August 1949). The life and works of Sir Archibald Garrod. *St Bartholomew's Hospital Journal*, **53**, 160–5, 186–90.

18. Letter from Robert S. Frew to C. J. R. Hart. Quoted in Hart, C. J. R. (August 1949). The life and works of Sir Archibald Garrod. *St Bartholomew's Hospital Journal*, **53**, 160–5, 186–90.

19. Ibid.

20. Ibid.

21. Garrod, A. E. (1899). Abernethian Society Lecture. Some clinical aspects of children's disease. *St Bartholomew's Hospital Journal*, **7**, 22–5.

22. Hippocrates. *The genuine works of Hippocrates* (trans. Francis Adams, LL.D.). Williams & Wilkins, Baltimore, 1939.

23. Many aspects of the life of Frederick Gowland Hopkins can be found in *Biographical memoirs of the Royal Society* (1972), Vol. 18.

24. Garrod, A. E. and Hopkins, F. G. (1896). On urobilin. Part I. The unity of urobilin. *Journal of Physiology*, **20**, 112–44; Garrod, A. E. and Hopkins, F. G. (1897–8). On urobilen. Part II. The percentage compostion of urobilen. *Journal of Physiology*, **22**, 451–64.

25. MacMunn, C. A. (1886). Researches on myohaematin and the histohaematins. *Philosophical Transactions*, **177**, 267–98.

26. Garrod, A. E. (1892). On the occurrence and detection of haematoporphyrin in the urine. *Journal of Physiology*, **13**, 598–620; Garrod, A. E. (1892). On the presence of uro-haemato-porphyrin in the urine in chorea and articulate rheumatism. *Lancet*, **i**, 793; Garrod, A. E. (1893). On haematoporphyrin as a urinary pigment in disease. *Journal of Pathology and Bacteriology*, **1**, 187–97; Garrod, A. E. (1894). A contribution to the study of the yellow colouring matter of the urine. *Proceedings of the Royal Society of London*, **55**, 394–407; Garrod, A. E. (1894). Some further observations on urinary haematoporphyrin. *Journal of Physiology*, **15**, 108–18; Garrod, A. E.

(1894–5). A contribution to the study of uroerythrin. *Journal of Physiology*, 17, 439–50; Garrod, A. E. (1894–5). Haematoporphyrin in normal urine. *Journal of Physiology*, 17, 349–52; Garrod, A. E. (1895) Late researches on urochrome. *Medical Press Circular*, NS 59, 238–40; Garrod, A. E. and Hopkins, F. G. (1896). Notes on the occurrence of large quantities of haematoporphyrin in the urine of patients taking sulphonal. *Journal of Pathology and Bacteriology*, 3, 435–48. *Transactions of the Pathological Society of London*, 47, 316–34; Garrod, A. E. and Hopkins, F. G. (1896). On urobilin. Part I. The unity of urobilin. *Journal of Physiology*, 20, 112–44; Garrod, A. E. (1897). Note on the origin of the yellow pigment of urine. *Journal of Physiology*, 21, 190–1; Garrod, A. E. and Hopkins, F. G. (1897–8). On urobilin. Part II. The percentage composition of urobilin. *Journal of Physiology*, 22, 451–64.

27. Garrod, A. E. (1897). The spectroscopic examination of urine. *Edinburgh Medical Journal*, NS 2, 105–16.
28. Garrod, A. E. (1893). On an unusual form of nodule upon the joints of the fingers. *St Bartholomew's Hospital Reports*, 29, 157–61. These were not the same nodules on the interphalangeal joints that occur in osteoarthritis. Those were described by Archibald's father, and are sometimes known as Garrod's pads. It seems likely that the nodules Garrod described were due to a variety of diseases and have no diagnostic significance.
29. Ibid.
30. Ord, W. M. and Garrod, A. E. (1895). *The climates and baths of Great Britain*. Report of a committee of the Royal Medical and Chirurgical Society of London. The Society, London.
31. Allbutt, T. C. (1897). *A system of medicine*. Macmillan, New York.
32. Letter from Dorothy Garrod to C. J. R. Hart. Quoted in Hart, C. J. R. The life and works of Sir Archibald Garrod, *St Bartholomew's Hospital Journal*, 53, 160–5, 186–90.

Chapter 5

1. Garrod, A. E. (1899). A contribution to the study of alkaptonuria. *Medical Chirurgical Transactions*, 82, 367–94.
2. Kuhn, T. S. (1962). *The structure of scientific revolutions*. University of Chicago Press.
3. Marcet, A. J. G. (1822). Account of a singular variety of urine, which turned black soon after being discharged. *Medical Chirurgical Transactions*, 12, 37. An interesting account of Marcet's life was written by Garrod (1925). Archibald Gaspard Marcet, Physician to Guy's Hospital, 1804–1819. *Guy's Hospital Reports*, 75, 373–87.
4. Bödeker, C. (1859). Über das Alkapton. *Zeitschrift für Rationelle Medizin*, 7, 130–45.
5. Knox, W. E. (1958). Sir Archibald Garrod's "Inborn Errors of Metabolism". II. Alkaptonuria. *American Journal of Human Genetics*, 10, 95–124.
6. Wolkow, M. and Baumann, C. (1891). Über das Wesen der Alkaptonuria. *Zeitschrift Physiologie Chemie*, 15, 228–85.

7. Bödeker, C. (1861). Das Alkapton; ein Beitrag zur Frage: welche Stoffe des Harns können aus einer alkalinischen Kupferoxydul reduciren? *Annals der Chemie Pharmacologie*, **117**, 98–106.
8. Virchow, R. (1866). Ein Fall von allgemeine Ochronose der Knorpel und knorpelähnlicher Theile. *Archiv für Pathologische Anatomie und Physiologie und für Klinische Medicin*, **37**, 212–19.
9. Osler, W. (1904). Ochronosis: the pigmentation of cartilages, sclerotics, and skin in alkaptonuria. *Lancet*, **i**, 10.
10. Garrod, A. E. (1898–9). Alkaptonuria: a simple method for the extraction of homogentisic acid. *Journal of Physiology*, **23**, 512–14.
11. Garrod, A. E. (1900). A contribution to the study of alkaptonuria. *Proceedings of the Royal Medical and Chirurgical Society*, NS **11**, 130–5.
12. Ibid., p. 133.
13. Ibid., pp. 133–4.
14. Ibid., p. 134.
15. Ibid., pp. 134–5.
16. Ibid., p. 135.
17. Osler, W. (1901). Letter to A. E. Garrod. Private and unpublished.
18. Quoted in Hart, C. J. R. (August 1949). The life and works of Sir Archibald Garrod. *St Bartholomew's Hospital Journal*, **53**, 160–5, 186–90.
19. Garrod, A. E. (1901). About alkaptonuria. *Lancet*, **ii**, 1484–6.
20. Garrod, A. E. (1902). About alkaptonuria. *Medico-Chirurgical Transactions*, **85**, 69–77.
21. Kirk, R. (1886). On a new acid found in human urine which darkens with alkalies. *British Medical Journal*, **2**, 1017–19.
22. Garrod (1901); (1902). *About alkaptonuria.*
23. Mittelbach, F. (1901). *Deutsches Archiv für Klinische Medicin*, **71**, 50.
24. Garrod (1902). *About alkaptonuria.*
25. A short discussion of the contribution of these scientists can be found in L. C. Dunn (1965). *A short history of genetics*, pp. 73–7. McGraw-Hill, New York.
26. Bateson, W. (1909). *Mendel's principles of heredity.* Cambridge University Press.
27. Bateson, W. and Saunders, E. R. (1901). *Report to the Evolution Committee of the Royal Society.* Footnote, Vol. 1, pp. 133–4.
28. Garrod, A. E. (11 January 1902). Letter to W. Bateson. Private and unpublished.
29. Ibid.
30. Bateson, W. and Saunders, E. R. (1981). *Report to the Evolution Committee of the Royal Society.*
31. Garrod, A. E. (11 January 1902). Letter to W. Bateson. Private and unpublished.
32. Garrod, A. E. (14 January 1902). Letter to W. Bateson. Private and unpublished.
33. Garrod, A. E. (11 January 1902). Letter to W. Bateson.
34. Garrod, A. E. (20 March 1902). Letter to W. Bateson. Private and unpublished.
35. Garrod, A. E. Letter to W. Bateson (18 June 1902). Private and unpublished.

36. Garrod, A. E. (1902). The incidence of alkaptonuria: a study in chemical individuality. *Lancet*, **ii**, 1616–20.
37. The engaging story, ascribed to Graham, that Garrod's concept of inborn errors of metabolism came as a sudden flash of insight as he was walking home from the Hospital for Sick Children at Great Ormond Street, should be accepted with caution. The evidence for this story does not hinge on documented or verifiable sources.

Chapter 6

1. Garrod, A. E. (1900). The Bradshaw Lecture on 'Urinary pigments in their pathological aspects.' Delivered before the Royal College of Physicians of London on 6 November. *Lancet*, **ii**, 1323–31.
2. Cammidge, P. J. and Garrod, A. E. (1900). On the excretion of diamines in cystinuria. *Journal of Pathology and Bacteriology*, **6**, 327–33.
3. Osler, W. (18 December 1903). Letter to A. E. Garrod. Private and unpublished.
4. Garrod, A. E. (5 June 1903). Letter to W. Bateson. Private and unpublished.
5. Huppert, C. H. (1896). Rectorial address delivered 16 November 1895. In *Die Erhaltung der Arteigenschaften*. Carl Ferdinand University Press, Prague. In his scholarly volume *Contrasts in scientific style*, published in 1990 by the American Philosophical Society, Philadelphia, Joseph Fruton wrote 'Hugo Huppert (1834–1904) had been a student of Carl Gotthelf Lehmann (1812–1918).' An interesting obituary of Huppert by R. von Zeynek was published in *Prager Medizinische Wochenschrift* (1904), **29**, 593–6.
6. Garrod, A. E. (1903). Über chemische Individualität und chemische Missbildungen. *Pflügers Archiv für die Gesamte Physiologie des Menschen und der Tiere*, **97**, 410–18.
7. Bateson, W. (1909). *Mendel's principles of heredity*, p. 227. Cambridge University Press. Bateson refers to the color of the urine in alkaptonuria as red in the first edition of his book. In 1913, alkaptonuric urine is still described as red (*Problems of genetics*, Cambridge University Press). In the fourth impression, in 1930 (Cambridge University Press) Bateson continues to assert that the urine is colored red in alkaptonurics. Bateson, of course, was not a physician; but it is none the less astonishing that this striking error was perpetuated.
8. Garrod, A. E. and Hele, T. S. (1905–6). The uniformity of the homogentisic acid excretion in alkaptonuria. *Journal of Physiology*, **33**, 198–205. Hele undertook this work while holding a research studentship at Bart's, where he had gone to pursue his clinical studies, graduating with a Bachelor of Medicine from the University of Cambridge in 1906. Stimulated by Garrod, he worked with Gowland Hopkins at Cambridge, but eventually chose an administrative career, becoming Master of Emmanuel College, Cambridge, and later Vice-Chancellor of the University.
9. Garrod, A. E., and Hurtley, W. H. (1905–6). On the estimation of homogentisic acid in urine by the method of Wolkow and Baumann. *Journal of Physiology*, **33**, 206–10.

10. Garrod, A. E. and Hurtley, W. H. (1906). Concerning cystinuria. *Journal of Physiology*, **34**, 217–23.
11. Garrod, A. E. and Hele, T. S. (1906–7). A further note on the uniformity of H:N quotient in cases of alkaptonuria. *Journal of Physiology*, **35**, 15–16.
12. Garrod, A. E. and Hurtley, W. H. (1907–8). On the supposed occurrence of uroleucic acid in the urine in some cases of cystinuria. *Journal of Physiology*, **36**, 136–41.
13. Garrod's application for the post of Assistant Physician at St Bartholomew's Hospital. January 1903. Privately printed.
14. Testimonial from F. H. Champneys, Physician-Accoucheur and Lecturer on Midwifery, St Bartholomew's Hospital, 15 December 1902.
15. Garrod, A. E. (1903–4). Lecture introductory to a course on chemical pathology. *St Bartholomew's Hospital Journal*, **11**, 20–2, 38–41.
16. Garrod, A. E. (1904). On black urine. *Practitioner*, **72**, 383–96.
17. Garrod, A. E. (1906–7). Abstract of a lecture on chorea. *St Bartholomew's Hospital Journal*, **14**, 88–9.
18. Garrod, A. E. (1903–4). Lecture introductory to a course on chemical pathology.
19. Ibid.
20. Punnett, R. C., *et al.* (1908). Mendelism in relation to disease. *Proceedings of the Royal Society of Medicine*, **8**, (i), 135–68.
21. Ibid., p. 160.
22. Ibid., pp. 161–2.
23. Ibid., p. 162.
24. Ibid., pp. 162–3.
25. Ibid., p. 164.
26. Ibid., pp. 164–6.
27. Ibid., p. 166.
28. Ibid., pp. 167–8.
29. Morgan, T. H., quoted in Beadle, G. W. (1963). *Genetics and modern biology: Jayne lectures.* American Philosophical Society, Philadelphia.
30. Gossage, A. M. (1908). The inheritance of certain human abnormalities. *Quarterly Journal of Medicine*, **1**, 331–44.
31. Gossage, A. M. (1913). Heredity. In *Diseases of children* (ed. A. E. Garrod, F. E. Batten, and H. Thursfield). Edward Arnold, London.
32. Bateson, W. (1906). An address on Mendelian heredity and its application to man. *Brain*, **29**, Part II, 157–79.
33. Harvey, R. D., archivist, John Innes Institute. Letter to A. G. Bearn (1 October 1990). Private and unpublished.
34. Cushing, H. (1940). *The life of Sir William Osler*, p. 735. Oxford University Press, Oxford.
35. The Association of Physicians of Great Britain and Ireland, Council minutes, 23 and 24 May 1908.
36. Reid, E. (1908). On ochronosis: report of a case; Osler, W., Clinical features; Garrod, A. E., The urine. *Quarterly Journal of Medicine*, **1**, 199–208.
37. The Association of Physicians of Great Britain and Ireland, Council minutes, 28 May 1928.

Chapter 7

1. Garrod, A. E. (1909). *Inborn errors of metabolism. The Croonian Lectures delivered before the Royal College of Physicians of London, in June, 1908*, p. 3. Oxford University Press, London.
2. Hunter, M. (1989). *Establishing the new science: the experience of the early Royal Society.* Boydell Press, Woodbridge, Suffolk.
3. Hartley, H. (1960). *The Royal Society: its origins and founders.* Royal Society, London.
4. Ibid.
5. Ibid.
6. Mrs Croone's will. Quoted in *College List of the Royal College of Physicians of London*, published annually.
7. Hartley, H. (1960). *Royal Society.*
8. Garrod, A. E. (1908). *The Croonian Lectures on inborn errors of metabolism.* Delivered before the Royal College of Physicians of London on 18, 23, 25, and 30 June 1908. *Lancet*, **ii**, 1–7, 73–9, 142–8, 214–20.
9. Unsigned review. (1909). *British Medical Journal*, **2**, 206.
10. Ibid.
11. Unsigned review. (1909). *Lancet*, **ii**, 22–3.
12. Halliburton, W. D., Hewitt, J. A., and Robson, W. (1936). *The essentials of chemical physiology: for the use of students*, (13th edn). Longmans Green, London; Halliburton, W. D. (1891). *A text-book of chemical physiology and pathology.* Longmans Green, London.
13. Halliburton, W. D. (1909). Review. *Nature*, **81**, 96.
14. Unsigned review (1909). *St Bartholomew's Hospital Journal*, **16**, 142.
15. Unsigned review (1909). *Journal of the American Medical Association*, **53**, 1427.
16. Garrod, A. E. (1909). Anomalies of urinary excretion. In *Modern medicine: its theory and practice*, Vol. 6, (ed. W. Osler and T. McCrae), pp. 40–85. Lea & Fébiger, Philadelphia; Garrod, A. E. (1909). Uraemia. In *Modern medicine: its theory and practice*, Vol. 6, pp. 86–102; Garrod, A. E. (1909). Uraemia or meningitis? *Proceedings of the Royal Society of Medicine*, **2**, (i), 169–75; Garrod, A. E. (1909–10). Concerning intermittent hydrarthrosis. *Quarterly Journal of Medicine*, **3**, 207–20; Garrod, A. E. (1909–10). Multiple peripheral neuritis in a child. *Proceedings of the Royal Society of Medicine*, **3**, (i), 38–40; Garrod, A. E. (1911). A case of spondylitis deformans. *Proceedings of the Royal Society of Medicine*, **4**, (i), 29–30; Garrod, A. E. (1911). On auscultation of the joints. *Proceedings of the Royal Society of Medicine*, **4**, (ii), 35–9; On auscultation of joints. *Lancet*, **i**, 213–14; Garrod, A. E. (1911). On the nature of the connection between erythemata and lesions of joints. *Lancet*, **i**, 1411–12.
17. Garrod, A. E. (1909). Enterogenous cyanosis. In *A system of medicine*, Vol. 5, (ed. C. Allbutt and H. D. Rolleston), pp. 838–45. Clarendon Press, Oxford.
18. Garrod, A. E. (1912). The Lettsomian Lectures on glycosuria. Delivered before the Medical Society of London. *Lancet*, **i**, 483–8, 562–77, 629–35.
19. Ibid.
20. Garrod, A. E. (1912–13). The scientific spirit in medicine: inaugural sessional address to the Abernethian Society. *St Bartholomew's Hospital Journal*, **20**, 19–27.

21. Ibid.

22. Ibid.

23. Garrod, A. E., Batten, F. E., and Thursfield, H. (ed.) (1913). *Diseases of children, by various authors*. Arnold, London. Garrod contributed the following sections: Disease as it affects children, pp. 1–4; Diseases of the ductless glands, pp. 560–84; Disorders of metabolism, pp. 585–602. The second edition, published in 1929, was edited H. Thursfield and D. Paterson; Garrod's section on metabolic disorders, pp. 530–57, was revised by E. A. Cockayne. The third edition, published in 1934, was also edited by Thursfield and Paterson; Garrod contributed a section on inborn errors of metabolism, pp. 583–92. The fourth edition, edited by D. Paterson and A. Moncrieff and published in 1947–9, was in two volumes, as was the fifth edition, edited by A. Moncrieff and P. Evans and published in 1953.

24. Unlike Garrod, however, Gossage seemed more interested in the pedigrees of a variety of malformations, and as early as 1907 he was writing about heredity in medicine in the *Quarterly Journal of Medicine*, perhaps at Garrod's urging.

25. Weil, A. (1884). Ueber die hereditäre Form des diabetes insipidus. *Archiv für Pathologische Anatomie und Physiologie und für Klinische Medicin*, **95**, 70–95.

26. Letters to Garrod (1915). Private and unpublished.

27. Garrod, A. E. (1890). Letter to J. Evans. Private and unpublished.

28. Cooke, A. M. (1964). *A history of the Royal College of Physicians of London*, Vol. 3, p. 966. Clarendon Press, Oxford.

29. Flexner, A. (1910). *Medical education in the United States and Canada*, Bull. No. 4, Carnegie Foundation for the Advancement of Teaching, New York.

30. *Report of the Royal Commission on University Education (1911–1913)*. Her Majesty's Stationery Office, London.

31. Garrod, A. E. (1914). Clinical applications of pathological chemistry. *Transactions of the 17th International Congress of Medicine, 1913*, pp. 71–81. Henry Froude, London. This was not, however, the first international congress that Garrod had attended. In 1900, when the Thirteenth Congress was held in Paris, Garrod was among those listed among the attenders, although he did not present a paper. His old chief, Sir Dyce Duckworth, was honorary president, and had probably encouraged Garrod to attend.

32. Ibid.

33. Hopkins, F. G. (1913). The dynamic side of biochemistry. *Nature*, **92**, 213–23.

Chapter 8

1. A general account of the medical services in the Mediterranean, including Malta, during the First World War can be found in Macpherson, W. G. (1921). Medical Services General History. In *History of the Great War based on official documents*, **1**, 235–48. His Majesty's Stationery Office, London. Details of the geographic distribution of the military hospitals and convalescent camps in the 17-mile-long, 8-mile-wide island are provided. Additional information relevant to the role of the British army in Malta can be found in Mackinson, A. G. (1916). *Malta — the nurse of the Mediterranean*. Hodder & Stoughton, London

2. Vella, F. (1965). Sir Archibald's stay in Malta 1915–1919. *St Bartholomew's Hospital Journal*, **69**, 138–45, and (1966). Sir Archibald Garrod and Malta: a historical occasion recalled. *St Luke's Hospital Gazette*, **1**, 41–50.

3. Garrod, A. E. (1914). Medicine from the chemical standpoint. *British Medical Journal*, **2**, 228–35.

4. The Order of St John was instituted in Jerusalem about 1085 as a community of monks who provided medical care for pilgrims. By the time they arrived in Malta, their devout heritage had been replaced by arrogance and a dismissal of the rights and privileges of the Maltese.

5. Vella (1965). 'Sir Archibald in Malta'.

6. Garrod, A. E. (1919). *Islands: a lecture delivered in the Aula Magna, Malta University on 21st January 1919*, p. 3. Empire Press, Malta.

7. Macpherson (1921). *History of the Great War*.

8. Letter from G. E. Graham to F. Vella, August 1964. Quoted in Vella (1966). 'Garrod and Malta', see ref. 2 above, p. 45.

9. Bruce, G. R. (n.d., *circa* 1919). *Military hospitals in Malta during the war*. Fortress Headquarters, Valletta, Malta, Quoted in Vella (1965). 'Sir Archibald in Malta'. *St Bartholomew's Hospital Journal*, (April 1965), **69**(4), 1–8.

10. Garrod, A. E. (1917). A variety of war heart which calls for treatment by complete rest. *Lancet*, **i**, 985–6.

11. Ibid.

12. Hurst, A. F. (1949). *A twentieth century physician: being the reminiscences of Sir Arthur Hurst, D.M., F.R.C.P.*, p. 143. Williams & Wilkins, Baltimore.

13. Hurst, A. F. (1927). The constitutional factor in disease. *British Medical Journal*, **1**, 823–7.

14. Zammit, T. (1929). *Malta: the islands and their history*, (2nd edn). Malta Herald, Valletta, Malta.

15. Zammit, T. and Singer, C. (1924). Neolithic representations of the human form from the islands of Malta and Gozo. *Journal of the Royal Anthropological Institute*, **54**, 67–100.

16. Garrod, A. E. (1918). *The University of Utopia*. Critiens Press, Malta.

17. Ibid.

18. Ibid.

19. Ibid.

20. Vella, F. (1965). Notes on Sir Archibald's stay in Malta 1915–1919. *St Bartholomew's Hospital Journal*, **69**, 138–45.

21. Haldane, J. B. S. (1928). *Possible worlds*. Chatto & Windus, London.

22. Garrod, A. N. (1916). Notes on the existence of a regimental M.O. *St Bartholomew's Hospital Journal*, **23**, 65–6.

23. Purves-Stewart, J. (1939). *Sands of time: recollections of a physician in peace and war*, p. 143. Hutchinson, London.

24. *Daily Malta Chronicle*, 28 December 1916. Quoted in Vella (1966). 'Garrod and Malta'. See ref. 2 above.

25. Bruce (n.d.). *Military hospitals in Malta*.

26. *Who was who 1929–1940*, Vol. 3, p. 935. Adam & Charles Black, London.

27. *Daily Malta Chronicle*, 16 January 1919, p. 4.

28. Ibid.

29. Ibid.

30. It has been estimated that approximately 300 000 individuals died in Germany during the influenza pandemic. The population in Germany in 1918 was approximately 60 million (Jordan, E. O. (1927). *Epidemic influenza: a survey.* American Medical Association, Chicago).

31. This chapter leans extremely heavily on the meticulous and scholarly research of Dr Frank Vella, Department of Biochemistry, University of Saskatchewan. Dr Vella, a medical graduate of The University of Malta, undertook extensive investigations in Malta to elucidate details about Garrod's stay there.

Chapter 9

1. Garrod, A. E. (1912). Statement handed to the Haldane Commission 9 November 1911. Appendix to *Fifth Report of The Commission. Minutes of Evidence October 1911 – January 1912, with appendices and index*, pp. 73–8. His Majesty's Stationery Office, London.

2. There was some question as to whether directors of professorial units should be full-time or part-time. Osler argued that directors should be part-time; Garrod believed this would be a serious mistake and would encourage mediocrity.

3. Garrod, A. E. (1919). The laboratory and the ward. In *Contributions to medical and biological research dedicated to Sir William Osler in honour of his seventieth birthday July 12, 1919, by his pupils and co-workers.* Vol. 1, pp. 59–69. Paul B. Hoeber, New York.

4. Ibid.

5. Graham, G. (1970). The formation of the medical and surgical professorial units in the London teaching hospitals. *Annals of Science*, **26**, 1–22.

6. Editorial (October 1919). *St Bartholomew's Hospital Journal*, **27**, 1.

7. Thomson, A. (9 January 1920). Letter to A. E. Garrod. Private and unpublished.

8. Herringham, W. P. (1936). Obituary. *Munk's Roll, 1826–1925*, Vol. 4, p. 334.

9. Thomson, A. (9 January 1920). Letter to A. E. Garrod.

10. Fraser, F. R. (1936). In Obituary of Archibald E. Garrod, *Lancet*, i, p. 808.

11. George, L. Letter to A. E. Garrod (21 February 1920). Private and unpublished.

12. Medicine at Oxford: Sir W. Osler's successor. *The Times*, 1 March 1920, p. 14.

13. Evans, E. (2 March 1920). Letter to A. E. Garrod. Private and unpublished.

14. Sir W. Osler's Successor. *The Times*, 3 March 1920, p. 16.

15. Garrod, A. E. (11 March 1920). Letter to the Dean of Christ Church. Private and unpublished.

16. The Regius Professorship of Medicine at Oxford was created by Henry VIII in 1546. The officeholder's sole responsibility was to supervise the education of the graduates and sanction their practice when they had arrived at a state of proficiency. The other disciplines, in addition to medicine, for which a Regius Professorship can be accorded are Divinity, Civil Law, Greek, and Hebrew. It was also in 1546 that Henry VIII renamed Cardinal College Christ Church

College. Regius Professorships at Cambridge were established by the King in 1540. Studentships at Christ Church correspond to 'Fellowships' or 'Scholarships' of other Oxford colleges, and are restricted to senior members.

17. Editorial notes (1920). *St Bartholomew's Hospital Journal*, **27**, 170.
18. Allbutt, T. C. (1920). Letter to A. E. Garrod. Private and unpublished.
19. Powell, N. (25 September 1920). Letter to A. E. Garrod. Private and unpublished.
20. Garrod had been head of the Professorial Medical Unit at Bart's for less than six months when he was approached by Oxford. He was appointed 13 August 1919, took office 1 October 1919, and resigned on his appointment as Regius Professor of Medicine on 30 September 1920. The recommendation that Garrod be appointed a Professor of Medicine in the University of London had not had time to be ratified by the University, and thus Garrod never officially held the title of Professor of Medicine in the University of London.
21. Garrod could have stayed on past the age of 70, since he was appointed before the age limit was introduced; but as he had been a member of the Royal Statutory Commission that put the limit into force, he felt he could not himself outstay his 70th birthday.
22. There is no official residence for the Regius Professor of Medicine. William Osler had made 13 Norham Gardens such as mecca for students, faculty, and visitors from overseas that it became known as 'The Open Arms'. The Garrods always lived on Banbury Road, first at 155, later at 133, and finally, in 1923, at number 85.

Chapter 10

1. Robb-Smith, A. H. T. (1970). *A short history of the Radcliffe Infirmary*. The Church Army Press, Oxford, for The United Oxford Hospitals; Gibson, A. G. (1926). *The Radcliffe Infirmary*. Humphrey Milford, Oxford University Press, London, pp. 260–80. An abbreviated history of the medical education at Oxford and Cambridge is provided by Robb-Smith, (1966) in his essay, 'Medical education in Britain prior to 1850'. In *The evolution of medical education in Britain*, (ed. F. N. L. Poynter), pp. 19–52. Williams & Wilkins, Baltimore. Oxford graduates frequently become disconsolate when they learn that Henry VIII had favored Cambridge over Oxford in creating a Regius Professorship there, albeit in Physic, six years earlier.
2. Haldane, J. B. S. (1920). Some recent work on heredity. *Transactions of the Oxford University Junior Science Club*, **1** (3), 3–11.
3. Haldane, J. B. S. (1958). The theory of evolution, before and after Bateson. *Journal of Genetics*, **56**, 1–17.
4. Garrod, A. E. (1923). Glimpses of the higher medicine. *Lancet*, **i**, 1091–6.
5. Cushing, H. (1940). *The life of Sir William Osler*. Oxford University Press, New York.
6. Survey report of the Medical School at the University of Oxford. In *Medical Education in England*, Vol. 2 (the provinces). Rockefeller Archive Center,

401A, Vol. 2. Pearce, Richard M. *Medical Education in England.* Vol. 2, copy 1, beginning p. 18, Box 23, Folder 301A. Note by Regius Professor of Medicine A. E. Garrod, January 1923.

7. Ibid.
8. Ibid.
9. Ibid.
10. Oxford University Hebdomadal Council Papers (1920–7), pp. 116–38.
11. Ibid.
12. Sisam, K. (1 November 1927). Memorandum. Archives Oxford University Press, P. P. 1780. As Sisam put it, 'I assented palely, and he agreed that it might come on the Agenda after delivery on November 24th. He will not pay for it. I don't see how we can very well avoid doing it. It ought not to cost more than £100, and might even sell a few copies.'
13. Oxford University Faculty Council. (1921). Minutes of Medical Faculty.
14. Ibid.
15. Ibid
16. Macbeth, R. (1968). The Radcliffe 40 years ago. *Oxford Medical School Gazette*, **20**, 139.
17. Macbeth, R. (2 August 1975). Letter to A. G. Bearn. Private and unpublished.
18. Macbeth (1968). *The Radcliffe 40 years ago.*
19. Krebs, H. (1970). Oxford medicine: essays on the evolution of the Oxford Clinical School to commemorate the bicentenary of the Radcliffe Infirmary 1770–1970 (ed. K. Dewhurst). Sandford Publications.
20. Ord, M. G., and Stocken, L. A. (1990). *The Oxford biochemistry department: its history and activities 1920–1985.* Joshua Associates, Oxford.
21. Pearce, R. M. (1923). Diary. Rockefeller Archive Center, Record Group 12.1 (Officers' Diaries).
22. Pearce, R. M. (1923). Report of Visit to Oxford to G. E. Vincent, Historical Record, University of Oxford, Biochemistry. Rockefeller Archive Center, Record Group RG 1.1, 401A, Oxford University-Biochemistry. Historical Record (p. 11) Box 21, Folder 279.
23. Peters, R. A. (16 February 1923). Letter to A. E. Garrod. Copy in Rockefeller Archive Center, Record Group RG 1.1, 401A, Oxford University-Biochemistry. Historical Record (pp. 14–16) Box 21, Folder 279. Minutes 2318.9, original filed with Cambridge University.
24. Pearce, R. M. (1923). Memorandum to A. Gregg and G. E. Vincent. Rockefeller Archive Center, Record Group RG 1.1, 401A, Oxford University. Historical Record (p. 19) Box 21, Folder 279. In the same memorandum, a request for £5000 to help build an extension for the Department of Anthropology received a laconic 'nothing doing'.
25. Garrod, A. E. (21 December 1923). Letter to E. R. Embree. Rockefeller Archive Center, Record Group RG 1.1, 401A, Oxford University. Historical Record Box 21, Folder 279.
26. Survey report of the Medical School at Oxford.
27. Hopkins, F. G. (17 July 1927). Letter to A. E. Garrod. Private and unpublished.

28. Drummond, J. C. (1924). Review of *Inborn errors of metabolism*. *Nature*, **113**, 595–6.
29. *Lancet* (1 September 1923), Vol. **205**, **ii**, p. 464.
30. Mackey, L. and Garrod, A. E. (1921). On congenital porphyrinuria, associated with hydroea aestivale and pink teeth. *Quarterly Journal of Medicine*, **15**, 319–30.
31. Garrod, A. E. (1936). Congenital porphyrinuria: a postscript. *Quarterly Journal of Medicine*, **29**, 473–80.
32. Garrod, A. E. (1909). *Inborn errors of metabolism*. Oxford University Press, Oxford.
33. Huppert, C. H. (1896). Rectorial address delivered 16 November 1895. In *Die Erhaltung der Arteigenschaften*, Prague. Carl Ferdinand University Press, Prague.
34. Garrod (1909). *Inborn errors*.
35. Ibid.
36. Garrod (1923). *Higher medicine*.
37. Ibid.
38. Ibid.
39. Huppert (1896). Rectorial address.
40. Garrod, A. E. (1924). The debt of science to medicine. *British Medical Journal*, **2**, 747–52.
41. Ibid.
42. Garrod, A. E. (1926). An address on the science of clinical medicine. *Lancet*, **ii**, 735–7.
43. Ibid.
44. Garrod, A. E. (5 October 1926). Letter to T. A. Malloch. Private and unpublished.
45. Hurst, C. C. (20 June 1931). Letter to Archibald E. Garrod. Private and unpublished.

Chapter 11

1. Oxford University Hebdomadal Council Minutes, 11 February 1928.
2. Oxford University Hebdomadal Council Minutes, 24 January 1927.
3. Simcock, A. V. (ed.) (1985). *Robert T. Gunther and the old Ashmolean*. Museum of the History of Science, Oxford.
4. Garrod, A. E. (1927). The Huxley Lecture on diathesis, delivered at the Charing Cross Hospital 24 November. *British Medical Journal*, **2**, 967–71; *Lancet*, **ii**, 1113–18.
5. Draper, G. (1924). *Human constitution: a consideration of its relationship to disease*. W. B. Saunders, Philadelphia.
6. Garrod (1927). Huxley lecture.
7. Hurst, A. F. (1927). An address on the constitutional factor in disease, delivered before the Ulster Medical Society 17 November 1926. *British Medical Journal*, **1**, 823–7, 866–8; *The constitutional factor in disease*, Kegan Paul, London.

8. Ibid.
9. Editorial (1927). *British Medical Journal*, **1**, 887–8.
10. Garrod, A. E. (1929). The power of personality. *British Medical Journal*, **2**, 509–12.
11. Garrod, A. E. (1927). Teaching of Clinical Medicine in England, In *Methods and problems of medical education*. **Eighth series**. The Rockefeller Foundation, New York.
12. Garrod, A. E. (1928). An address on the place of biochemistry in medicine, delivered at the opening of the Courtauld Institute of Biochemistry at the Middlesex Hospital 14 June. *British Medical Journal*, **1**, 1099–101.
13. Ibid.
14. Poynton, F. J. (1927). Compilation of reminiscences in School of Nursing Library, Hospital for Sick Children, Great Ormond Street, London, after retirement of Archibald E. Garrod. Unpublished personal documents.
15. Garrod, A. E. (1928). The lessons of rare maladies, the annual oration delivered before the Medical Society of London on 21 May. *Lancet*, **i**, 1055–60.
16. The handsome house where the Garrods lived at 10 Chandos Street was eventually torn down, and an unattractive office building has been erected on the site.
17. Garrod (1928). *Lessons of rare maladies.*
18. Garrod, A. E. (1929). In memoriam. Sir Dyce Duckworth, Bart., M.D., 1840–1928. *St Bartholomew's Hospital Reports*, **62**, 18–30.
19. Garrod, A. E. (1929). J. F. Bullar, correspondence, *St Bartholomew's Hospital Journal*, **36**, 110.
20. Garrod, A. E. (1931). *The inborn factors in disease: an essay.* Clarendon Press, Oxford.
21. Archives, Oxford University Press. Diathesis: Garrod, Sir Archibald, 1 November 1927, K.S., P.P. 1780.
22. Garrod, A. E. (1931). *The inborn factors in disease: an essay.*
23. Barlow, T. (3 November 1932). Letter to A. E. Garrod. Private and unpublished.
24. Unsigned review (1936). *British Medical Journal*, **1**, 731–3.
25. Unsigned review (1932). *Lancet*, **i**, 939.
26. Unsigned review (1931). *Journal of the American Medical Association*, **97**, 1174.
27. Unsigned review (1932). *New England Journal of Medicine*, **206**, 1016.
28. Unsigned review, *Lancet*, (1932).
29. Report on meeting of the Historical Section of the Royal Society of Medicine. (1932). *British Medical Journal*, **1**, 523–4.
30. Ibid.
31. Garrod, A. E. (1933). Chemistry and medicine: summary of lecture delivered to pre-clinical students. *St Bartholomew's Hospital Journal*, **41**, 27–8.
32. On 16 June 1932 Gowland Hopkins opened a discussion on recent advances in the study of enzymes and their action. *Proceedings of the Royal Society of London*, **61B**, 280–97. Garrod, as Vice-President of the Royal Society, probably attended this lecture.

33. Penrose, L. S. (1934). *The influence of heredity on disease*, Buckston Browne Prize Essay, 1933. H. K. Lewis, London.
34. Lenz, F. (1919). Die Bedeutung der statistisch ermittelten Belastung mit Blutverwandtschaft der Eltern. *Münchener Medizin Wochenschrift*, **66**, 1340.
35. Garrod, A. E. (1903). Über chemische Individualität und chemische Missbildungen. *Pflügers Archiv für die gesamte Physiologie des Menschen und der Tiere*, **97**, 410–18.
36. Garrod, A. E. (4 April 1934). Letter to L. S. Penrose. Private and unpublished.
37. Garrod is probably referring here to the paper by L. Hogben, R. L. Worrall, and I. Zieve (1932). *Proceedings of the Royal Society of Edinburgh*, **52**, 264–95, in which Hogben reviews the inheritance patterns of patients with alkaptonuria and gives Garrod great credit; and also to Hogben, L. (1933), *Nature and nurture*, W. W. Norton, New York, where Hogben writes 'His (Garrod's) 1902 paper is a landmark in the history of human genetics.'
38. Garrod, A. E. Letter to E. A. Cockayne (22 March 1933). Private and unpublished.
39. Penrose, L. S. (6 April 1934). Letter to A. E. Garrod. Private and unpublished.
40. Ibid.
41. Penrose, L. S. (23 August 1935). Letter to A. E. Garrod. Private and unpublished.
42. Fölling, A. (1934). Über Ausscheidung von Phenylbrenztraubensäure in den Harn als Stoffwechselanomalie in Verbindung mit Imbezillität. *Hoppe-Seyler Zeitschrift für Physiologische Chemie*, **227**, 169–76. Three years later, this condition was named phenylketonuria by Penrose and Quastel: Penrose, L. and Quastel, J. H. (1937). Metabolic studies in phenylketonuria. *Biochemical Journal*, **31**, 266–74.
43. Fölling, A. (19 August 1935). Letter to A. E. Garrod. Private and unpublished.
44. Mackey, L. and Garrod, A. E. (1922). On congenital porphyrinuria, associated with hydroa aestivale and pink teeth. *Quarterly Journal of Medicine*, **15**, 319–30; Mackey, L. and Garrod, A. E. (1926). A further contribution to the study of congenital porphyrinuria (haematoporphyria congenita). *Quarterly Journal of Medicine*, **19**, 357–73.
45. Garrod, A. E. (1936). Congenital porphyrinuria. A postscript. *Quarterly Journal of Medicine*, **29**, 473–80. The clinical features suggest that Garrod's patient was suffering, as he suspected, from congenital erythropoietic porphyria, which is now thought to be due to a decrease in the enzyme hydroxymethylbilane hydroxylase. The disease is indeed rarer than alkaptonuria, and Garrod was correct in anticipating a greater frequency of consanguinity.
46. Obituary, Sir Archibald E. Garrod. *The Times*, 30 March 1936.
47. Hopkins, F. G. (1936). Sir Archibald Garrod, K.C.M.G., F.R.S. *Nature*, **137**, 770–1.
48. Hopkins, F. G. (1938). Archibald Edward Garrod (1857–1936). *Obituary Notices of the Royal Society of London*, **6**, 225–8.
49. Hopkins, F. G. (1936). Obituary, Sir Archibald Garrod. *British Medical Journal*, **1**, 775–6.

50. Thursfield, H. (1936). Obituary, Sir Archibald Edward Garrod, K.C.M.G., M.D., F.R.S. *Lancet*, **i**, 807–9.
51. Dryden, J. (1950). Prologue to the University of Oxford. In *The poetical works of John Dryden*, new edition revised and enlarged by George R. Noyes. Riverside Press, Cambridge. The 'Prologue' was written in 1676 and was first published in *Miscellany poems*, 1684; Obituary, Sir Archibald Edward Garrod. (1936) *Lancet*, **i**, 807
52. Fraser, F. (1936). Obituary, A. E. Garrod. *Lancet*, **i**, p. 808
53. Obituary, Sir Archibald Garrod. (1936). *Journal of the American Medical Association*, **106**, 1830.
54. Royal Faculty of Physicians and Surgeons of Glasgow University. Excerpt from the minutes of a Meeting of Faculty held on 6 April 1936.
55. Graham, G. (1936). Obituary, Sir Archibald E. Garrod, K.C.M.G., D.M., F.R.C.P., F.R.S., 1857–1936. *St Bartholomew's Hospital Reports*, **69**, 12–26.
56. Report on the memorial service for A. E. Garrod. *The Times*, 4 April 1936, p. 18.
57. Garrod made his last will, witnessed by his long-standing friend Sir Humphrey Rolleston, on 12 February 1932. He left an estate of over £19 000, an amount that enabled Laura Garrod to live comfortably in Cambridge, close to Dorothy and a few friends. Four years later, on 21 February 1940, Laura died of a coronary thrombosis in her 77th year. She and Archibald were buried in the graveyard of the parish church at Melton, near the graves of their three sons.

   On 18 December 1968, Dorothy, the Garrods' remaining child, died; her ashes are buried in the churchyard alongside her parents and her three brothers. Dorothy never married, so that, from now on, the tradition of academic excellence so evident in the Garrod family must be continued by collateral relatives.
58. Yeo, G. (1990). *Ward names at Bart's*. The Royal Hospital of St Bartholomew, London.

Chapter 12

1. Beadle, G. W. (11 December 1958). Genes and chemical reactions in *Neurospora*. In *Nobel Lectures including presentation speeches and laureates' biographies. Physiology or medicine 1942–1962*, pp. 587–97. Elsevier, Amsterdam.
2. Ibid.
3. Sapp, J. (1990). Founding father fables. In *Where the truth lies: Franz Moewus and the origins of molecular biology*, pp. 27–55. Cambridge University Press.
4. Scriver, C. R. and Childs, B. (1989). *Garrod's inborn factors in disease*. Oxford University Press.
5. Sapp (1990). *Where the truth lies*. In this series of essays, Sapp discusses the work of Garrod as an example of a founding-father fable. He also analyzes the concept of a precursor. Sapp concludes that 'the casting of Archibald Garrod as the "father" of biochemical genetics was done by Beadle to supply his experimental evidence for the one gene–one enzyme theory with a plot that would parallel the

rediscovery of Mendel's work'. See also Sapp, J. (1983). The struggle for authority in the field of heredity. *Journal of the History of Biology*, **16**, 311–42.

6. Beadle, G. W. (1966). Biochemical genetics: some recollections. In *Phage and the origins of molecular biology*, (ed. J. Cairns, G. S. Stent, and J. D. Watson), pp. 23–32. Cold Spring Harbor Laboratory of Quantitative Biology, Cold Spring Harbor, New York; Beadle, G. W. (1945). Biochemical genetics. *Chemical Reviews*, **37**, 15–96.

7. Beadle, G. W. (1945). The genetic control of biochemical reactions. *Harvey Lectures*, **40**, 179–94.

8. Beadle, G. W., and Tatum, E. L. (1941). Experimental control of development and differentiation: genetic control of developmental reactions. *American Naturalist*, **75**, 107–16.

9. Bateson, W. (1909). *Mendel's principles of heredity*, p. 266. Cambridge University Press.

10. Wagner, R. P. (1989). On the origins of the gene–enzyme hypothesis. *Journal of Heredity (Brief Communications)*, **80**, 503–4.

11. Cuénot, L. (1903). L'heredité de la pigmentation chez les souris (2me note). *Archives de Zoologie Experimentale et Generale*, 4e sér., **1**(3), 33–41.

12. Bateson, W. (1909). *Mendel's principles*; idem (1913) *Problems of genetics*. Yale University Press, New Haven.

13. Haldane, J. B. S. (1954). *The biochemistry of genetics*. Allen & Unwin, London.

14. Bateson, W. (1906). An address on Mendelian heredity and its application to man. *Brain*, **29**, Part II, 157–79.

15. Sutton, W. S. (1903). The chromosomes in heredity. *Biological Bulletin*, **4**, 231–51.

16. Bateson, W., Saunders, E. R., Punnett, R. C., and Hurst, C. C. (1905). *Reports to the Evolution Committee of the Royal Society*, Report 2. Harrison & Sons, London.

17. Bateson, W. (1928). *Essays and addresses*, p. 93. Cambridge University Press.

18. Bateson (1909). *Mendel's principles*.

19. Bateson (1913). *Problems of genetics*.

20. Bateson (1909). *Mendel's principles*, p. 254.

21. Lederberg, J. (1990). Edward Lawrie Tatum, December 14, 1909–November 7, 1975. *Biographical Memoirs*, Vol. 59, p. 357. National Academy Press, Washington, DC.

22. Hardy, G. H. (1908). Mendelian proportions in a mixed population. *Science*, **28**, 49–50.

23. Punnett, R. C., *et al.* (1907–8). Mendelism in relation to disease. *Proceedings of the Royal Society of Medicine*, **8**, i, 135–68; Yule's remarks are on pages 164–6.

24. Weinberg, W. (1908). Über dem Nachweis der Vererbung beim Menschen. *Jahreshelfte, Verein für Vaterländische Naturkunde in Württemberg*, **64**, 369–82. English translation in Boyer, S. H. (1963). *Papers on human genetics*, Prentice-Hall, Englewood Cliffs, New Jersey.

25. Gross, O. (1914). Über den Einfluss des Blutserums des normalen und des alkaptonurikens auf Homogentisinsäure. *Biochemische Zeitschrift*, **61**, 165–70.

26. La Du, B. N., Zannoni, V. G., Laster, L., and Seegmiller, J. E. (1958). The

nature of the defect in tyrosine metabolism in alcaptonuria. *Journal of Biological Chemistry*, **230**, 251–60.

27. Zannoni, V. G., Seegmiller, J. E., and La Du, B. N. (1962). Nature of the defect in alcaptonuria. *Nature*, **193**, 952–3.

28. Wheldale, M. (1907). The inheritance of flower colour in *Antirrhinum majus. Proceedings of the Royal Society of London*, **79B**, 288–305.

29. Wheldale, M. (1916). *The anthocyanin pigments of plants.* Cambridge University Press.

30. Haldane, J. B. S. (1920). Some recent work on heredity. *Transactions of the Oxford University Junior Science Club*, **1** (3), 3–11.

31. From Rose Scott-Moncrieff's valuable account of the classical period in chemical genetics in *Notes and Records of the Royal Society of London*, (1981), **36**, 126–54.

32. Scott-Moncrieff, R. (1937). The biochemistry of flower colour variation. In *Perspectives in biochemistry*, (ed. J. Needham and D. E. Green), pp. 230–43. Cambridge University Press.

33. Morgan, T. H. (1926). Genetics and the physiology of development. *American Naturalist*, **60**, 489–515.

34. Morgan, T. H. (4 June 1934). The relation of genetics to physiology and medicine. In *Nobel Lectures including presentation speeches and laureates' biographies. Physiology or medicine 1922–1941*, pp. 313–28, Elsevier, Amsterdam. The presentation speech was delivered by F. Henschen, Professor, Royal Caroline Institute (p. 312), who emphasized more than Morgan the importance of genetics in human disease:

> ... considering the present attitude of medicine and the dominating place of the constitutional researches, the role of the inner, hereditary factors as to health and disease appears in a still clearer light. For the general understanding of maladies, for prophylactic medicine, and for the treatment of diseases, hereditary research thus gains still greater importance.

35. Wright, S. (7 May 1980). Letter to A. G. Bearn. Private and unpublished.

36. Ibid.

37. Haldane, J. B. S. and Huxley, J. S. (1927). *Animal biology.* Clarendon Press, Oxford.

38. Wells, H. G., Huxley, J. S., and Wells, G. P. (1937). *The science of life.* Doubleday, Doran, New York. Originally published in England in 1929.

39. Goldschmidt, R. B. (1938). *Physiological genetics.* McGraw-Hill, New York; idem (1930). Theory of the gene, *Scientific Monthly*, **48**, 268–73.

40. Lederberg (1990). *Edward Lawrie Tatum.*

41. Morgan (1934). *Relation of genetics to physiology and medicine*, pp. 313–28.

42. Hopkins, G. (1898). The chemistry of the urine. In *Textbook of physiology*, (ed. E. A. Schäfer), Vol. 1, pp. 570–639. Young J. Pentland, Edinburgh; Macmillan, New York.

43. Bauer, J. H. (1924). *Die konstitutionelle Disposition zu inneren Krankheiten*, (3rd edn). Verlag von Julius Springer, Berlin; Hueppe, F. (1893). Die Ursachen der Gärungen und Infektionskrankheiten und deren Beziehungen zum Kausalproblem und zur Energetik. *Verhandlungen der Gesellschaft*

Deutscher Naturforscher und Ärzte; Martius, F. (1909). *Pathogenesse innerer Krankheiten: nach Vorlesungen für studierende und Ärzte*. Franz Deuticke, Leipzig and Vienna.

44. Adami, J. G. (1907). Inheritance and disease. In *Modern medicine: its theory and practice*, (ed. W. Osler, assisted by T. McCrae), Vol. 1, pp. 17–50. Lea Brothers, Philadelphia.
45. Ibid., pp. 639–47.
46. Bodansky, M. (1927). *Introduction to physiological chemistry*. Wiley, New York.
47. Bodansky, M. (1934). *Introduction to physiological chemistry*, (3rd edn). Wiley, New York.
48. Williams, R. J. (1931). *An introduction to biochemistry*. Van Nostrand, New York.
49. Williams, R. J. (1956). *Biochemical individuality: the basis for the genetotrophic concept*. Wiley, New York.
50. Haldane, J. B. S. (1937). The biochemistry of the individual. In *Perspectives in biology: thirty-one essays presented to Sir Gowland Hopkins by past and present members of his laboratory*, (ed. J. Needham and D. E. Green), pp. 1–10. Cambridge University Press.
51. Ibid.
52. Ibid.
53. Beadle, G. W. (8 November 1974). Letter to A. G. Bearn. Private and unpublished.
54. Lederberg (1990). *Edward Lawrie Tatum*, p. 366.
55. Haldane, J. B. S. (1942). *New paths in genetics*. Harper, New York.
56. Müller, H. J. (1947). Pilgrim Trust Lecture: The gene. *Proceedings of the Royal Society of London*, **B134**, 1–37.
57. Haldane, (1942). *New paths*.
58. Beadle, G. W. (20 November 1974). Letter to A. G. Bearn. Private and unpublished.
59. Beadle, G. W., and Tatum, E. L. (1941). Experimental control of development and differentiation: genetic control of developmental reactions. *American Naturalist*, **75**, 107–16.
60. Beadle, G. W. (1939). Genetic control of the production and utilization of hormones. *Proceedings of the Seventh International Genetical Congress*, (ed. R. C. Punnett), pp. 58–61. Cambridge University Press.
61. Bonner, D. H. (1946). Biochemical mutations in *Neurospora*. *Cold Spring Harbor Symposia on Quantitative Biology*, **11**, 14–24.
62. Delbrück, M. (1946). Discussion following Bonner's presentation. *Cold Spring Harbor Symposia on Quantitative Biology*, **11**, 22–23.
63. Lederberg, J. (1956). Comments on gene–enzyme relationship. In *Enzymes: units of biological structure and function*, Henry Ford Hospital International Symposia, (ed. O. H. Gaebler), pp. 161–74. Academic Press, New York.
64. Horowitz, N. H., and Leupold, U. (1951). Some recent studies bearing on the one gene–one enzyme hypothesis. *Cold Spring Harbor Symposia on Quantitative Biology*, **16**, 65–74.
65. Bonner, D. M. (1951). Gene–enzyme relationships in *Neurospora*. *Cold Spring Harbor Symposia on Quantitative Biology*, **16**, 143–53.

66. Horowitz, N. H. (1991). Fifty years ago: the *Neurospora* revolution. *Genetics*, **127**, 631–5.

67. In discussing the initial failure of medical scientists to appreciate the relation of genetics to human medicine, Beadle wrote [Beadle, G. W. (8 November 1974). Letter to A. G. Bearn. Private and unpublished]: 'I believe progress in human genetics was slowed greatly by both the belief that man was an almost impossible organism for genetic study and the skepticism of many M.D.s as to the significance of genetics.' 'I was an advisor,' Beadle continued, 'to W. H. Freeman, when Curt Stern's book on human genetics was conceived, written, and published. (In 1949) I urged the publisher to publicize it widely among medical practitioners — that every M.D. should have a copy. (A sample mailing was sent out with a response so small that it scarcely paid the postage on the mailing.)' [Stern, C. (1949). *Principles of human genetics*. W. H. Freeman, San Francisco.]

68. Childs, B. (1982). Genetics in the medical curriculum. *American Journal of Human Genetics*, **13**, 319–24; Childs, B. (1989). Acceptance of the Howland Award. *Pediatric Research*, **26**, 390–3.

Chapter 13

1. Delbrück, M. (1946). Discussion following Bonner's presentation. *Cold Spring Harbor Symposia on Quantitative Biology*, **11**, 555–6.

2. Lederberg, J. (1956). Comments on gene–enzyme relationship. In *Enzymes: units of biological structure and function*, Henry Ford Hospital International Symposium, (ed. O. H. Gaebler), pp. 161–74. Academic Press, New York.

3. It is indeed fortunate that it has been recently republished with important and illuminating commentaries by Charles Scriver and Barton Childs. Scriver, C. R., and Childs, B. (1989). *Garrod's inborn factors in disease*. Oxford University Press.

4. Hirs, C. H. W., Moore, S., and Stein, W. H. (1960). The sequence of the amino acid residues in performic acid-oxidized ribonuclease. *Journal of Biological Chemistry*, **235**, 633–47.

5. Butterfield, H. (1957). *The origins of modern science* (2nd edn, revised). Macmillan, New York.

6. Hacking, I. (1983). *Representing and intervening: introductory topics in the philosophy of natural science*. Cambridge University Press.

7. Ibid.

8. For a contemporary account of inborn errors of metablism, see Scriver, C. R., Beaudet, A. L., Sly, W. S., and Valle, D. (ed.) (1989). *The metabolic basis of inherited disease*, (6th edn). McGraw-Hill Information Services, New York.

9. Milch, R. A. (1960). Studies of alcaptonuria: inheritance of 47 cases in eight highly inter-related Dominican kindreds. *American Society of Human Genetics*, **12**, 76–85.

10. Ingram, V. M. (1956). A specific chemical difference between the globins of normal human and sickle-cell anaemia haemoglobin. *Nature*, **178**, 792–4.

11. McKusick, V. A. (1992). *Mendelian inheritance in man. Catalogs of autosomal dominant, autosomal recessive, and X-linked phenotypes*. 10th edn. Johns

Hopkins University Press. (Online version, OMIM accessible from Welch Medical Library, Johns Hopkins University.)

12. Scriver, C. R., Beaudet, A. L., Sly, W. S., and Valle, D. (ed.) (1989). *The metabolic basis of inherited disease.*

13. King, R. A., Rotter, J. L., and Motulsky, A. G. (ed.) (1992). *The genetic basis of common diseases.* Oxford University Press.

14. Harris, H. (1969). Enzyme and protein polymorphism in human populations. In 'New aspects of human genetics'. *British Medical Bulletin*, **25**, 5.

15. Lewontin, R. C. and Hubby, J. L. (1966). A molecular approach to the study of genetic heterozygosity in natural populations. II. Amount of variation and degree of heterozygosity in natural population of *Drosophila pseudoobscura*. *Genetics*, **54**, 595.

16. Skamene, E. (1991). Population and molecular genetics of susceptibility to tuberculosis. *Clinical and Investigative Medicine*, **14**(2), 160–6.

17. Committee on Mapping and Sequencing the Human Genome. (1988). *Mapping and sequencing the human genome.* National Research Council. National Academy Press, Washington, DC.

APPENDIX I

◆

Sir Archibald Garrod's Antecedents

TRADITION in the Garrod family holds that they are descended from a Huguenot refugee from Flanders or France who settled in England, but opinions differ whether the founding father migrated in the sixteenth century or at about the time of the Revocation of the Edict of Nantes by Louis XIV in 1685. There were substantial numbers of Huguenot refugees in the counties of Norfolk, Suffolk, Essex, and Cambridgeshire from the latter part of the sixteenth century, and they were prominent in Norwich, Ipswich, and other towns, particularly those with long-established textile industries. However, the name Garrod is unlikely to be of Huguenot origin.[1] It has been a common name in Suffolk for the last two centuries, certainly, and in Stradbroke, Suffolk, the parish registers of baptisms, marriages, and burials show that there were at least half-a-dozen households of this name living there in the latter part of the eighteenth century, some of whose members were illiterates or paupers receiving parish relief. Among these Stradbroke Garrods were Robert and his wife Hannah, the earliest ancestors of Sir Archibald Garrod who have been positively identified through the parish records in Suffolk.

Robert and Hannah had nine children baptized at Stradbroke between 1759 and 1775, but the parish registers do not record the baptism of their eldest son, also named Robert. It may be inferred, therefore, that Robert and Hannah moved to Stradbroke from another village, probably in the 1750s. Robert was a tenant farmer of fairly modest means: when he died, in 1809, the value of his personal property was certified as above £600 but under £800. Hannah received under his will an annuity of £17, a bed, and other furniture (the annuity was about the same as the annual earnings of a farm laborer). When their son Robert died, in 1835, his personal estate was sworn as being worth under £450. At the time of the 1851 census, Robert's son, another Robert, was shown as farming 120 acres at Buttlesea Green, Stradbroke, and employing four men. At that time, there were some

5000 farms in Suffolk, 1600 of which comprised 150 acres or more; thus, Robert Garrod's holding, as a tenant, was very much in the middle range.

James, the second son of Robert and Hannah, was baptized at Stradbroke in 1761, but at the time of his marriage to Catherine Borrett in 1783 he lived in Redgrave, a few miles away. After his marriage, he returned to Stradbroke to farm, and the parish register records the baptisms of seven children between 1787 and 1799, including four sons, the second of whom was yet another Robert (baptized in 1793). James followed the family tradition of tenant farming in Stradbroke.

Robert Garrod (1793–1877)

With this Robert Garrod, the son of James, the family took a new direction. Hitherto they had been middle-ranking tenant farmers, but now they produced three generations of people of the highest academic and intellectual achievement, including three Fellows of the Royal Society and the first woman to hold a Chair at Cambridge University.

Robert, himself from humble beginnings, achieved a high place in the business and political life of Ipswich. At the age of thirteen he left his home at Stradbroke to take an apprenticeship with Trotman's Brewery in Ipswich. Four years later, through the influence of Captain Studd, a partner in the brewery who was also an employee of the East India Company, he took a leave of absence to sail to India on the company's ship, *Carnatic*.[2] Robert returned to the brewery after a year; the business failed in 1820. After being involved in its liquidation, Robert entered into a partnership with John King of Stowmarket, who combined an auctioneer's and estate agent's business (with a branch in Ipswich) with printing and publishing. Two years after King died, his son decided to concentrate on the printing business; the partnership was dissolved, and Robert Garrod set himself up in business in 1833 on his own account as an auctioneer and estate agent in the Buttermarket in Ipswich. This business flourished (under the name Garrod & Turner after 1852, when Robert took William Turner into partnership), and remained a separate concern until it was amalgamated with another estate agent's business in the 1980s.[3]

Ipswich was a rapidly expanding town in the mid-nineteenth century, and Robert would have been well placed to profit from its development, although he had plenty of competition from other estate agents and auctioneers in Ipswich and nearby market towns. The population of Ipswich increased from 20 000 in 1831 to 43 000 in 1871. The wet dock — said to be the largest in England — was opened in 1842, and the town was connected by rail to Colchester and Bury St Edmunds in 1846, and to Norwich in 1849. Engineering and the manufacture of artificial fertilizers were the main industries.

Robert Garrod's progress in business and social stature can be traced through his descriptions of himself in official documents. When he married Sarah Ennew Clamp in 1816, he was listed in the parish marriage register as a brewer's clerk; in the 1832 poll book — the first to record him as having a vote in Parliamentary elections — he is a printer; in the 1841 and 1851 census returns he is an auctioneer; but in 1861 he is 'auctioneer, estate agent and joint occupier of 195 acres'; in the 1871 census he is an auctioneer and landowner. By 1876, Robert Garrod owned his substantial house in Lower Brook Street, Ipswich, together with the Golden Lion Hotel and four smaller properties in the town. He had also acquired a 65-acre farm and a house (later known as Wilford Lodge) in the village of Melton near Woodbridge, another substantial house in the village, and two small cottages. When he died, in 1877, his estate was valued at just under £45 000. His land and houses passed to his son, Sir Alfred Baring Garrod.

Robert Garrod was active in local politics in the Whig interest, and was said to have played a part in the bitterly contested 1820 Parliamentary election in Ipswich. In 1866 he was chairman of the committee that successfully secured the election of two Whig candidates for the first time in many years. Also in 1866, when he was seventy-three years old, he was elected an alderman by the Ipswich Council. When Robert Garrod died, it was reported that he had been a generous benefactor of the East Suffolk Hospital and had taken a considerable interest in its welfare.

Robert and his wife, Sarah, had one son and four daughters. Through their children's marriages they became connected with the prominent Ipswich families of Colchester, Turner, and Sparrow, and with Charles Keene, the artist and cartoonist of *Punch*, and Meredith Townsend of the *Spectator*.

Their son was given the name Alfred Baring. There is a tradition in the family that he was named Baring because of a promise made to a Garrod, presumably James, by a member of the Baring family. James had rescued this Baring from highwaymen, and in gratitude, Baring is said to have promised to finance the legal education of one of James's sons, provided this son bestowed the name of Baring on his son. No evidence of this episode has been found, nor does it seem that any of James's sons was educated in the law. Another possible explanation is that Alfred, who was born in 1819, was given his second name in honor of Henry Baring, who was a last-minute Reform candidate for the Ipswich constituency in the 1818 Parliamentary election, and as such would have enjoyed the support of Alfred's father. Baring came close to being elected, and was given a hero's reception in the town after the election. As was invariably the case at this time, the election was bitterly contested, and there were accusations of violence and corruption on both sides. The adulation of Henry Baring lasted well into 1819.

Alfred Baring Garrod was educated at Ipswich Grammar School, which at that time offered no formal teaching of science; the curriculum concentrated on Latin and Greek, as was the practice in the major public schools and grammar schools. After he left school, he was apprenticed to Charles Chambers Hammond, a physician at the East Suffolk Hospital in Ipswich. Hammond was the son of a close neighbor of the Garrods: his father was active in local politics in the Whig interest, and he himself became an alderman of Ipswich in later life. Alfred Baring Garrod completed his medical education at University College Hospital, London.

In 1845, Alfred Baring Garrod married Elizabeth Ann Colchester, whose family was prominent in business in Ipswich, and through this marriage became related to the Sparrow family, who were active in both the political and the business life of Ipswich. There were six children of the marriage, of whom two sons, Alfred Henry and Archibald Edward, both became Fellows of the Royal Society.

Notes

This account of the Garrod family is based on records in the Ipswich branch of the Suffolk Record Office, including parish registers of baptisms, marriages, and burials; parish rate books; census enumerators' returns; wills; poll books; and newspapers.

The prime source for the life of Robert Garrod — Sir Archibald Garrod's grandfather — is the obituary in the *East Anglian Daily Times* of 12 June 1877.

1. Opinion from the Huguenot Society of Great Britain and Ireland. The obituary of Robert Garrod in the *East Anglian Daily Times* of 12 June 1877 contains information that could only have been given to a close friend, but no mention is made of the Huguenot connection, of which Robert Garrod might reasonably have been proud in view of the prominence of families of Huguenot descent in Ipswich.
2. East India Company ships were being built at Ipswich at this time.
3. William Turner married Garrod's daughter Rosa in 1855. The Turner family remained in the auction and estate agent business until recent times.

APPENDIX II

Chronology

AEG stands for Archibald Edward Garrod.

1792   Birth of Robert Garrod, AEG's grandfather (d. 1887).
1819   Birth of Alfred Baring Garrod, FRS, father of AEG (d. 1907), distinguished physician and rheumatologist. Birth of Elizabeth Ann Colchester, mother of AEG (d. 1891).
1845   Birth of Charles Robert, the first of AEG's four siblings (d. 1862). Died of tuberculosis at age 17.
1846   Birth of Alfred Henry, FRS, AEG's brother (d. 1879), distinguished zoologist. Died of tuberculosis at age 33.
1849   Birth of Herbert Baring, AEG's brother (d. 1912), barrister and classical scholar.
1852   Birth of Edith Kate, AEG's sister (d. 1924).
1857   25 November, birth of Archibald Edward Garrod at 84 Harley Street, London.
1864   Birth of Laura Elisabeth Smith, AEG's future wife (d. 1940).
1867   The *Handbook of classical architecture* hints at an early scholarly bent. Birth of Helen Effie Garrod, AEG's favorite sister (d. 1899). Died of tuberculosis at age 32.
1869   AEG becomes interested in natural history and writes letters to his parents about butterflies.
        AEG writes *The tiger*, a schoolboy foray into science.
1873   AEG enters Marlborough College (Littlefield House) in Wiltshire.
1874   A. B. Garrod and family move to 10 Harley Street.
1875   AEG leaves Marlborough, having won the Royal Geographical Society's bronze medal for physical geography.
1876   AEG receives Certificate in Chemistry from University College, London.
1877   AEG enters Christ Church, Oxford, as a Commoner.
1879   AEG wins Johnson Memorial Prize on an astronomical subject.
1880   AEG achieves a First in Natural Science and earns a BA in June. He enters St Bartholomew's Hospital, London.
1881   AEG wins a Junior Scholarship at Bart's.
1882   AEG's *The nebulae: a fragment of astronomical history* is published.
1884   AEG wins the Brackenbury Scholarship in Medicine at Bart's. During the summer, AEG travels to Norway with Henry Lewis Jones, and visits the leper colony in Bergen. AEG delivers a paper on leprosy to the Abernethian Society.

1885   AEG earns degrees of Bachelor of Medicine and Surgery.
       Travels to Vienna to spend the winter months at the Allgemeines
       Krankenhaus.
1886   AEG and Laura Elisabeth Smith are married.
       AEG publishes *An introduction to the use of the laryngoscope*.
       Receives Doctorate of Medicine from Oxford.
       Elected to Royal Medical and Chirurgical Society.
1887   Birth of the Garrods' first son, Alfred Noël.
       AEG holds post of Casualty Physician at Bart's.
       Alfred Garrod's Knighthood.
1888   AEG appointed Assistant Physician to the West London Hospital.
       Elected Fellow of the Royal College of Physicians (FRCP).
1890   AEG publishes *A treatise on rheumatism and rheumatoid arthritis*.
1892   AEG appointed Assistant Physician at the Hospital for Sick Children,
       Great Ormond Street.
       Birth of Dorothy Annie Elizabeth Garrod, CBE (d. 1968), distinguished
       archaeologist.
       AEG appointed Assistant Physician to Out-Patients of the Hospital for
       Sick Children.
1894   AEG elected Member of the Physiological Society.
       Birth of Thomas Martin Garrod (d. 1915).
       AEG became Physician to the Alexander Hospital for Diseases of the Hip.
1895   AEG elected Medical Registrar and Demonstrator in Morbid Anatomy at
       Bart's.
1896   AEG publishes first paper jointly with F. Gowland Hopkins.
1897   Birth of Basil Rahere Garrod (d. 1919).
1899   AEG appointed Physician at Great Ormond Street.
       Delivers a lecture to the Abernethian Society, 'Some clinical aspects of
       children's disease'.
       Publishes 'A contribution to the study of alkaptonuria' in *Medico-
       Chirurgical Transactions*.
1900   AEG delivers the Bradshaw Lecture, *Urinary pigments in their
       pathological aspects*, to the Royal College of Physicians.
1901   AEG publishes paper entitled 'The incidence of alkaptonuria: a study in
       chemical individuality' in the *Lancet*.
1903   AEG appointed Assistant Physician at Bart's.
1907   AEG becomes joint editor of the *Quarterly Journal of Medicine*.
1908   AEG delivers the Croonian Lectures, *Inborn errors of metabolism*, before
       the Royal College of Physicians.
1909   AEG becomes a Councillor, Royal College of Physicians.
       Inborn errors of metabolism published.
1910   AEG elected a Fellow of the Royal Society.
1912   AEG becomes Full Physician at Bart's, age 55.
       Delivers the Lettsomian Lectures on glycosuria before the Medical
       Society of London.
1913   AEG edits *Diseases of children*.

1914   AEG speaks before the Haldane Commission (the Royal Commission on University Education).
        LL. D. (*honoris causa*) University of Aberdeen.
        War is declared.
1915   Thomas Martin Garrod, Lieutenant of the Loyal North Lancashire Regiment, is killed in action.
        AEG is posted to Malta.
1916   Alfred Noël Garrod, Lieutenant, Royal Army Medical Corps, is killed in action.
        MD (*honoris causa*), University of Malta.
1918   KCMG conferred upon AEG.
1919   AEG is made Director of the Medical Unit at Bart's.
1920   AEG becomes Regius Professor of Medicine at Oxford.
        MD (*honoris causa*), University of Dublin.
        Delivers the Schörstein Lecture, *Diagnosis of diseases of the pancreas*, at the London Hospital Medical College.
        AEG made Consulting Physician and Governor of St Bartholomew's Hospital.
1922   MD (*honoris causa*), University of Padua.
1923   AEG delivers the Linacre Lecture, 'Glimpses of the Higher Medicine', at Cambridge.
        Member of Medical Research Council 1923–8.
        LL. D. (*honoris causa*), University of Glasgow.
1924   AEG delivers the Harveian Oration on the debt of science to medicine before the Royal College of Physicians of London.
1925   AEG receives Osler Memorial Medal.
1926   AEG becomes Vice-President of the Royal Society.
1927   AEG elected to the Royal Society Club.
        Retires from Oxford at the age of 70.
        Delivers the Huxley Lecture 'On diathesis'.
1928   AEG delivers the Annual Oration before the Medical Society of London.
1931   AEG publishes *The inborn factors in disease*, his final work.
1935   AEG receives the Gold Medal of the Royal Society of Medicine.
1936   28 March, Archibald Edward Garrod dies in Cambridge.

APPENDIX III

Additional Notes and References

Chapter 1

King, L. S. (1991). *Transformations in American medicine: from Benjamin Rush to William Osler.* The Johns Hopkins University Press, Baltimore.
Webster, C. (ed.). (1981). *Biology, medicine and society 1840–1940.* Cambridge University Press.
West, R. (1982). *1900.* Viking Press, New York.

Chapter 3

Moore, N. (1918). *The history of St Bartholomew's Hospital*, Vol. 2. C. Arthur Pearson, London.
Power, D'A., and Waring, H. J., (1923). *A short history of St Bartholomew's Hospital 1123–1923.* St Bartholomew's Hospital, London.
Poynter, F. N. L. (1966). *The evolution of medical education in England.* Williams & Wilkins, Baltimore.

Chapter 4

Billroth, T. (1924). *The medical sciences in the German universities.* Macmillan, New York.
Cabot, A. T. (1897). Science in medicine. *Boston Medical and Surgical Journal,* **137**, 481–4.
Cochrane, R. G. (1959). *Leprosy in theory and practice.* Wright, Bristol.
Conrad, J. (1885). *The German universities for the last fifty years,* (trans. John Hutchison). David Bryce & Son, Glasgow.
Landsteiner, K. (1900). Zur Kenntnis der antifermentativen, lytischen und agglutinierenden Wirkungen des Blutserums und der Lymphe. *Zentralblatt für Bakteriologie,* **27**, 357–62.
Neuberger, M. (1943). *British medicine and the Vienna School: contacts and parallels.* Heinemann Medical, London.

Chapter 6

Provine, W. B. (1971). *The origins of theoretical population biology.* University of Chicago Press.

Chapter 7

Knox, W. E. (1958). Sir Archibald Garrod's "Inborn Errors of Metabolism". I. Cystinuria. *American Journal of Human Genetics*, **10**, 3–32.

Knox, W. E. (1958). Sir Archibald Garrod's "Inborn Errors of Metabolism". III. Albinism. *American Journal of Human Genetics*, **10**, 249–267.

Knox, W. E. (1958). Sir Archibald Garrod's "Inborn Errors of Metabolism". IV. Pentosuria. *American Journal of Human Genetics*, **10**, 385–97.

Chapter 8

Evans, J. D. (1971). *The prehistoric antiquities of the Maltese islands: a survey.* Athlone Press, University of London.

Chapters 9 and 10

Allbutt, C. (1920). The universities in medical research and practice. *British Medical Journal*, **2**, 1–8.

Allbutt, C. (1921). Some comments on clinical units. *Lancet*, **ii**, 937–40.

Edmund B. Ford, professor of genetics at Oxford, in correspondence with the author, remembered Garrod at this time as being 'a rather heavily-built man, a little above middle height, with a somewhat red, full, face and abundant curly white hair.' Letter to A. G. Bearn, 1978. Private and unpublished.

Harris, H. (1988). Medical research in South Parks Road 1920–1987. In *A county hospital 1920–1988: a collection of essays about the history, changes, and development of the hospitals in Oxford*, produced by the Oxfordshire Health Authority for the exhibition 'The Growth of a County Hospital 1920–1988'. British Association for the Advancement of Science, Oxford.

Peters, R. A. (1930). Department of Biochemistry, University of Oxford. *Methods and Problems of Medical Education*, **18**, 109–18.

Robb-Smith, A. H. T. (1968). The development and future of the Oxford medical school. *Transactions of the Society for Occupational Medicine*, **18**, 13–21.

Chapter 11

Bauer, J. (1924). *Die konstitutionelle Disposition zu inneren Krankheiten*, (3rd edn). Springer, Berlin.

Draper, W. H. (1888). On the relation of scientific to practical medicine. *Transactions of the Association of American Physicians*, **3**, 1–8.

Harcourt, A. V. (1910). The Oxford museum and its founders. *Cornhill Magazine*, **28**, (no. 165), 350–63.

Chapter 12

Abir-Am, P. G. (1985). Themes, genres and orders of legitimation in the consolidation of new scientific disciplines: deconstructing the historiography of molecular biology. *History of Science*, **23**, 73–117.

Cook, R. (1937). A chronology of genetics. In *US Department of Agriculture Yearbook*, pp. 1457–77.

Cuénot, L. (1902). La loi de Mendel et l'hérédité de la pigmentation chez les souris. *Archives de Zoologie Experimentale et Générale*, 3e sér., **10**, 27–30.

Dronamraju, K. R. (1989). *The foundations of human genetics*. Charles C. Thomas, Springfield, Illinois.

Ephrussi, B. (1942). Chemistry of "eye color hormones" of *Drosophila*. *Quarterly Review of Biology*, **17**, 327–38.

Haldane, J. B. S. (1965). *Enzymes*. MIT Press, Cambridge, Mass.

Kay, L. E. (1989). Selling pure science in wartime: the biochemical genetics of G. W. Beadle. *Journal of the History of Biology*, **22**, 73–101.

Muller, H. J. (1922). Variation due to change in the individual gene. *American Naturalist*, **56**, 32–50.

Needham, J. and Baldwin, E. (eds) (1949). *Hopkins and biochemistry 1861–1947: papers concerning Sir Frederick Gowland Hopkins, O.M., P.R.S., with a selection of his addresses and bibliography of his publications*. W. Heffer and Sons, Cambridge.

Provine, W.B. (1986). *Sewall Wright and evolutionary biology*. University of Chicago Press.

Rimington, C. (1990). History of haem biosynthesis, porphyrins and porphyrias seen in retrospect. *Molecular Aspects of Medicine*, **11**, 7–10.

Sapp, J. (1987). *Beyond the gene: cytoplasmic inheritance and the struggle for authority in genetics*. Oxford University Press, New York and Oxford.

Stent, G. S. (1972). Prematurity and uniqueness in scientific discovery. *Scientific American*, **227**, 84–93.

Williams, G. C. (1991). The dawn of Darwinian medicine. *Quarterly Review of Biology*, **66**, 1–22.

Zuckerman, H. and Lederberg, J. (1986). Postmature scientific discovery? *Nature*, **324**, 629–31.

Chapter 13

Friedmann, T. (1991). Molecular medicine. In *The genetic revolution*, (ed. B. D. Davis), pp. 132–51. Johns Hopkins University Press, Baltimore.

Needham, J. and Baldwin, E. (ed.) (1949). *Hopkins and biochemistry 1861–1947*. W. Heffer and Sons, Cambridge.

Miscellaneous

Bearn, A. G. (1975–6). Lettsomian Lectures I. Archibald Garrod and the birth of an idea. *Transactions of the Medical Society of London*, **92**, 47–56.

Bearn, A. G. (1975–6). Lettsomian Lectures II. Present concepts and future directions. *Transactions of the Medical Society of London*, **92**, 57–63.

Bearn, A. G., and Miller, E. D. (1979). Archibald Garrod and the development of the concept of inborn errors of metabolism. *Bulletin of the History of Medicine*, **53**, 315–28.

Castle, W. E. (communicated 1919). Piebald rats and the theory of genes. *National Academy of Sciences*, **5**, 126–30.

Childs, B. (1970). Sir Archibald Garrod's conception of chemical individuality: a modern appreciation. *New England Journal of Medicine*, **282**, 71–7.

Childs, B. (1981). Genetic factors in human disease. In *Genetic issues in pediatric and obstetric practice*, (ed. J. Kaback). Year Book Medical Publishers, Chicago.

Childs, B. (1987). Genetics in medical education. *American Journal of Human Genetics*, **41**, 296–303.

Coleman, W. (1977). *Biology in the nineteenth century*. Cambridge University Press.

Crowther, J. G. (1952). *British scientists of the twentieth century*. Routledge & Kegan Paul, London.

Darden, L. (1991). *Theory change in science: strategies from Mendelian genetics*. Oxford University Press.

Garrod, S. C. (1989). Family influences on A. E. Garrod's thinking. *Journal of Inherited Metabolic Diseases*, **12** (Suppl. 1), 2–8.

Glass, B. (1965). A century of biochemical genetics. *Proceedings of the American Philosophical Society*, **109**, 227–36.

Glass, B. (1974). The long neglect of genetic discoveries and the criterion of prematurity. *Journal of the History of Biology*, 7, 101–4.

Glass, B. (1988). *A guide to the genetics collections of the American Philosophical Society*. American Philosophical Society Library, Philadelphia.

Goldstein, J. L. (1986). On the origins of PAIDS (paralyzed academic investigator's disease syndrome). *Journal of Clinical Investigation*, **78**, 848–54.

Haldane, J. B. S. (1908). Opening address to the British Association. The relation of physiology to physics and chemistry. *Nature*, **78**, 553–6.

Harris, H. (1963). *Garrod's 'Inborn errors of metabolism'*. Oxford University Press, London.

Harris, H. (1970). *The principles of human biochemical genetics*. North-Holland, Amsterdam.

Hibbert, C. (ed.) (1988). *The encyclopaedia of Oxford*. Macmillan, London.

Holmes, F. L. (1986). Patterns of scientific creativity. *Bulletin of the History of Medicine*, **60**, 19–35.

Kevles, D. J. (1981). Genetics in the United States and Great Britain 1890–1930: a review with speculations. In *Biology, medicine and society*, (ed. C. Webster), pp. 193–215. Cambridge University Press.

Kohler, R. E. (1982). From medical chemistry to biochemistry: the making of a biomedical discipline. Cambridge University Press.

Olby, R. (1974). *The path to the double helix*. University of Washington Press, Seattle.

Peters, R. A. (1958). Some reminiscences of Sir Archibald E. Garrod, K.C.M.G., F.R.C.P., F.R.S. *The American Journal of Human Genetics*, **10**, 1–2.

Appendix III

Rutz, C. (1970). The Garrods. In *Zurich Papers on Medical History*, New Series No. 77, (ed. Prof. E. H. Ackerknecht). Juris Druck und Verlag, Zurich.

Sandler, I. and Sandler, L. (1986). On the origin of Mendelian genetics. *American Zoologist*, **26**, 753–68.

Shine, I. and Wrobel, S. (1976). *Thomas Hunt Morgan: pioneer of genetics*. University Press of Kentucky, Lexington.

Skamene, E. (1991). Population and molecular genetics of susceptibility to tuberculosis. *Clinical and Investigative Medicine*, **14**, 160–6.

Stephens, A. D. (1989). Cystinuria and its treatment: 25 years experience at St Bartholomew's Hospital. *Journal of Inherited Metabolic Diseases*, **12**, 197–209.

Temkin, O. (1977). The scientific approach to disease: specific entity and individual sickness. In *The double face of Janus*, pp. 441–55. Johns Hopkins University Press, Baltimore.

Wells, H. G. (1925). *Chemical pathology; being a discussion of general pathology from the standpoint of the chemical processes involved*, (5th edn). W. B. Saunders, Philadelphia and London.

APPENDIX IV

◆

Bibliography of the Writings of Archibald E. Garrod

Based on a list (pp. 20–6) appended to an obituary notice by George Graham in *St Bart's Hosp. Rep.*, **69**, pp. 12–19 (1936).

1882

The nebulae: a fragment of astronomical history. Parker, Oxford, 1882. (Johnson Memorial Prize Essay of 1879; reconstructed and rewritten, 1881.)

1884

A visit to the leper hospital at Bergen (Norway). (Abstract.) *St Bart's Hosp. Rep.*, **30**, pp. 311–13 (1884).

1885

Some cases of sclerosis of the spinal cord. *St Bart's Hosp. Rep.*, **21**, pp. 93–9 (1885).

1886

A case of paralysis of the abductors of the vocal cords, with lesions of several cranial nerves. *St Bart's Hosp. Rep.*, **22**, pp. 209–11 (1886).
An introduction to the use of the laryngoscope. Longmans Green, London, 1886.
Hysteria. (Abstract). *St Bart's Hosp. Rep.*, **22**, p. 364 (1886).

1887

Rutpure of trachea, followed by general emphysema and asphyxia. *Med. Press. Circ.*, N.S. **45**, p. 519 (1887).

1887–8

A contribution to the theory of the nervous origin of rheumatoid arthritis. (Abstract.) *Proc. roy. med. chir. Soc. Lond.*, N.S. **2**, pp. 310–13 (1887–8).
A further contribution to the study of rheumatoid arthritis. (Abstract.). *Proc. roy. med. chir. Soc. Lond.*, N.S. **2**, pp. 372–6 (1887–8).

1888

A contribution to the theory of the nervous origin of rheumatoid arthritis. *Med.-chir. Trans.*, 2nd ser., **53**, pp. 89–105 (1888).
A further contribution to the study of rheumatoid arthritis. *Med.-chir. Trans.*, 2nd ser., **53**, pp. 265–81 (1888).
On the relation of erythema multiforme and erythema nodosum to rheumatism. *St Bart's Hosp. Rep.*, **24**, pp. 43–54 (1888).
(Archibald E. Garrod and E. Hunt Cooke.) An attempt to determine the frequency of rheumatic family histories amongst non-rheumatic patients. *Lancet*, **ii**, p. 110 (1888).

1889

On the relation of chorea to rheumatism, with observations of eighty cases of chorea. *Med.-chir. Trans.*, **72**, pp., 145–63 (1889).

The pathology of chorea: a suggestion. *Lancet*, **ii**, p. 1051 (1889).

1890

A treatise on rheumatism and rheumatoid arthritis, Griffin, London, 1890. Translated into French as *Traité du rhumatisme et de l'arthrite rhumatoïde. Traduit par le Dr. Brachet*, Paris (1891).

1891

A case of gouty periostitis. *Lancet*, **ii**, pp. 1334–5 (1891).

Notes on the common haemic cardiac murmur. *St Bart's Hosp. Rep.*, **27**, pp. 33–40 (1891).

1892

The changes in the blood in the course of rheumatic attacks. *Med.-chir. Trans.*, **75**, pp. 189–225 (1892).

On the occurrence and detection of haematoporphyrin in the urine. *J. Physiol.*, **13**, pp. 598–620 (1892).

On the presence of uro-haemato-porphyrin in the urine in chorea and articulate rheumatism. *Lancet*, **i**, p. 793 (1892).

1893

On an unusual form of nodule upon the joints of the fingers. *St Bart's Hosp. Rep.*, **29**, pp. 157–61 (1893).

On haematoporphyrin as a urinary pigment in disease. *J. Path. Bact.*, **1**, pp. 187–97 (1893).

1894

A contribution to the study of the yellow colouring matter of the urine. *Proc. roy. Soc.*, **55**, pp. 394–407 (1894).

On the association of cardiac malfunction with other congenital defects. *St Bart's Hosp. Rep.*, **30**, pp. 53–61 (1894).

Some further observations on urinary haematoporphyrin. *J. Physiol.*, **15**, pp. 108–18 (1894).

(W. P. Herringham, A. E. Garrod, and W. J. Gow.) *A handbook of medical pathology. For the use of students in the Museum of St Bartholomew's Hospital*, Baillière, Tindall and Cox, London, 1894.

1894–5

A contribution to the study of uroerythrin. *J. Physiol.*, **17**, pp. 439–50 (1894–5).

Haematoporphyrin in normal urine. *J. Physiol.*, **17**, 349–52 (1894–5). How to look up a point in a medical library. *St Bart's Hosp. J.*, **2**, pp. 145–6 (1894–5).

1895

Arthritis deformans. In *Twentieth century practice: an international encyclopaedia of modern medical science*, Vol. 2 (ed. Thomas L. Stedman), pp. 511–74 (1895).

Case of a cretin under thyroid treatment. *Trans. med. Soc. Lond.*, **18**, pp. 368–9 (1895).

A case of sclerema neonatorum ending in recovery. *Lancet*, i, pp. 1103–5 (1895).

A case of sclerema neonatorum ending in recovery. *Trans. Med. Soc. Lond.*, **18**, pp. 314–23 (1895).

Late researches on urochrome. *Med. Press Circ.*, N.S. **59**, pp. 238–40 (1895). The medicinal springs of Great Britain. Bath (W. M. Ord and A. E. Garrod), pp. 515–27; Buxton (W. M. Ord and A. E. Garrod), pp. 528–36; Matlock Bath, pp. 537–8; Droitwich, pp. 561–6; Nantwich, pp. 567–9; Leamington, pp. 583–8; Cheltenham, pp. 589–92; Tunbridge Wells, pp. 593–5. In *The climates and baths of Great Britain. Being the report of a committee of the Royal Medical and Chirurgical Society of London. ...* Vol. I (1895).

A specimen of urine rendered green by indigo. *Trans. clin. Soc. Lond.*, **28**, pp. 307–9 (1895).

(A. E. Garrod and H. Morley Fletcher.) The maternal factors in the causation of rickets. *Brit. med. J.*, ii, pp. 707–11 (1895).

(Sir Dyce Duckworth and A. E. Garrod.) A case of hepatic cirrhosis, with obstruction in the superior vena cava. *St Bart's Hosp. Rep.*, **32**, pp. 71–7 (1895).

1896

On the pigmentation of uric acid crystals deposited from urine. *J. Path. Bact.*, **3**, pp. 100–6 (1896).

The rationale of the accepted treatment of gout. *Med. Press Circ.*, N.S. **62**, pp. 227–30 (1896).

(A. E. Garrod and F. Gowland Hopkins.) Notes on the occurrence of large quantities of haematoporphyrin in the urine of patients taking sulphonal. *J. Path. Bact.*, **3**, pp. 435–48 (1986); *Trans. Path. Soc. Lond.*, **47**, pp. 316–34 (1896).

(A. E. Garrod and F. Gowland Hopkins.) On urobilin. Part I. The unity of urobilin. *J. Physiol.*, **20**, pp. 112–44 (1896).

(Translation.) (1896). *A treatise on cholelithiasis*. By B. Naunyn. Translated by Archibald E. Garrod, New Sydenham Society, London.

1897

A case of sclerema neonatorum. *Trans. clin. Soc. Lond.*, **30**, pp. 129–32 (1897). Chronic rheumatism, (pp. 56–60); Muscular rheumatism, (pp. 61–4); Gonorrhoeal rheumatism, (pp. 64–72); Rheumatoid arthritis (in part), (pp. 73–102). In, *A system of medicine*, Vol. 3 (ed. Sir Thomas Clifford Allbutt), (1897).

Malformation of the aortic valves; ulcerative endocarditis; associated malformation of the liver. *Trans. path. Soc. Lond.*, **48**, pp. 42–5 (1897).

Note on the origin of the yellow pigment of urine. *J. Physiol.*, **21**, pp. 190–1 (1897).

The spectroscopic examination of urine. *Edinb. med. J.*, N.S. **2**, pp. 105–16 (1897).

Über den Nachweis des Hämatoporphyrins im Harn. *Cbl. inn. Med.*, **18**, pp. 497–9 (1897).

(A. E. Garrod, A. A. Kanthack, and J. H. Drysdale.) On the green stools of typhoid fever, with some remarks on green stools in general. *St Bart's Hosp. Rep.*, **33**, pp. 13–23 (1897).

[Translation.] A contribution to the clinical and bacteriological study of the Brazilian framboesia or 'boubas'. By Achilles Breda. Translated by Archibald E. Garrod. In *New Sydenham Society: Selected essays and monographs*, pp. 259–83 (1897).

[Translation.] On polypapilloma tropicum (framboesia). By M. Charlouis. Translated by Archibald E. Garrod. In *New Sydenham Society: Selected essays and monographs*, pp. 285–319 (1897).

1897–8

(F. Gowland Hopkins and A. E. Garrod.) On urobilin. Part II. The percentage composition of urobilin. *J. Physiol.*, **22**, pp. 451–64, (1897–8).

1898

Carcinoma of the oesophagus which proved fatal by perforation of the aorta. *Trans. path. Soc. Lond.*, **49**, pp. 92–3 (1898).

A case of achondroplasia. *Trans. clin. Soc. Lond.*, **31**, pp. 294–5 (1898).

1898–9

Alkaptonuria: a simple method for the extraction of homogentisinic acid from the urine. *J. Physiol.* **23**, pp. 512–14 (1898–9).

1899

A contribution to the study of alkaptonuria. *Med.-chir. Trans.*, **82**, pp. 369–94 (1899); *Proc. roy. med. chir. Soc.*, N.S. **II**, pp. 130–5 (1899).

1899–1900

Some clinical aspects of children's disease. An address delivered before the Abernethian Society, November 9th 1899. *St Bart's Hosp. J.*, 7, pp. 22–5 (1899–1900).

1900

The Bradshaw Lecture on the urinary pigments in their pathological aspects. Delivered before the Royal College of Physicians of London on Nov. 6, 1900. *Lancet*, **ii**, pp. 1323–31 (1900).

(Cammidge, P. J. and A. E. Garrod.) On the excretion of diamines in cystinuria. *J. Path. Bact.*, **6**, pp. 327–33 (1900).

[Translation.] A contribution to the aetiology of tertiary syphilis, with special reference to the influence of mercurial treatment upon the development of tertiary symptoms. By Thomas v. Marchalko. Translated by Archibald Garrod. In *New Sydenham Society: Selected essays and monographs (from foreign sources)*, pp. 1–53 (1900).

[Translation.] Contribution to the study of visceral affections in the early stages of syphilis. I. Icterus syphiliticus precox. By O. Lasch. ... Translated by Dr. Garrod. In *New Sydenham Society: Selected essays and monographs (from foreign sources)*, pp. 143–64 (1900).

1901

About alkaptonuria. *Lancet*, **ii**, pp. 1484–6 (1901); *Med.-Chir. Trans.*, **85**, pp. 69–77 (1902).

A case of haematoporphyrinuria not due to sulphonal. By James Calvert. With a report on the urine by A. E. Garrod. *Trans. clin. Soc. Lond.*, **34**, pp. 43–5 (1901).

The clinical and pathological relations of the chronic rheumatic and rheumatoid affections to acute infective rheumatism. *Lancet*, **i**, pp. 774–7 (1901). ˙

(Sir Dyce Duckworth and A. E. Garrod.) A contribution to the study of intestinal sand, with notes on a case in which it was passed. *Med.-chir. Trans.*, **84**, pp. 389–404 (1901).

1901–2
(K. J. P. Orton and A. E. Garrod.) The benzoylation of alkapton urine. *J. Physiol.*, **27**, pp. 89–94 (1901–2).

1902
The diagnostic value of melanuria. *St Bart's Hosp. Rep.*, **38**, pp. 25–32 (1902). Ein Beitrag zur Kenntnis der kongenitalen Alkaptonurie. *Cbl. inn. Med.*, **23**, pp. 41–4 (1902).

The incidence of alkaptonuria: a study in chemical individuality. *Lancet*, **ii**, pp. 1616–20 (1902).

(Sir Dyce Duckworth and A. E. Garrod.) A contribution to the study of intestinal sand, with notes on a case in which it was passed. *Lancet*, **i**, pp. 653–6 (1902).

1903
The diagnostic value of melanuria. *St Bart's Hosp. Rep.*, **38**, pp. 25–32 (1903).

Some further observations on the reaction of urochrome with acetaldehyde. *J. Physiol.*, **29**, pp. 335–40 (1903).

Ueber chemische Individualität und chemische Missbildungen. *Pflüg. Arch. ges. Physiol.*, **97**, pp. 410–18 (1903).

1903–4
Lecture introductory to a course on chemical pathology. *St Bart's Hosp. J.*, **ii**, pp. 20–2, 38–41 (1903–4).

1904
Concerning pads upon the finger joints and their clinical relationships. *Brit. med. J.*, **ii**, p. 8 (1904).

On black urine. *Practitioner*, **72**, pp. 383–96 (1904).

A survey of the recorded cases of haematoporphyrinuria not due to sulphonal. *Trans. path. Soc. Lond.*, **55**, pp. 142–51 (1904).

Tumour of the liver in a boy aet. 10 years. *Trans. clin. Soc. Lond.*, **37**, pp. 222–3 (1904).

1904–5
A clinical lecture on chorea. Delivered at the Hospital for Sick Children, Great Ormond Street, WC. *Clin J.*, **25**, pp. 257–63 (1904–5).

1905
Clinical diagnosis: the bacteriological, chemical, and microscopical evidences of disease. By Rudolf v. Jaksch. Fifth English edition. ... (Ed. by Archibald E. Garrod), Griffin, London, 1905.

Haematuria. *Trans. med. Soc. Lond.*, **28**, pp. 132–51 (1905).

(A. E. Garrod and Ll. Wynne Davies.) On a group of associated congenital malformations, including almost complete absence of the muscles of the abdominal wall, and abnormalities of the genito-urinary apparatus. *Med.-chir. Trans.*, **88**, pp. 362–81 (1905).

1905–6
(A. E. Garrod and T. Shirley Hele.) The uniformity of the homogentisic acid excretion in alkaptonuria. *J. Physiol.*, **33**, pp. 198–205 (1905–6).

(A. E. Garrod and W. H. Hurtley.) On the estimation of homogentisic acid in urine by the method of Wolkow and Baumann. *J. Physiol.*, **33**, pp. 206–10 (1905–6).

1906

Peculiar pigmentation of the skin in an infant. *Trans. clin. Soc. Lond.*, **39**, p. 216 (106).

Rheumatoid arthritis. *Practitioner*, **76**, pp. 376–87 (1906).

(A. E. Garrod and W. H. Hurtley.) Concerning cystinuria. *J. Physiol.*, **34**, pp. 217–23 (1906).

(A. E. Garrod and F. Langmead.) A case of associated congenital malformations, including transposition of viscera. *Trans. clin. Soc. Lond.*, **39**, pp. 131–5 (1906).

(A. E. Garrod and F. J. Steward.) A case or primary pneumococcal peritonitis. *Lancet*, **ii**, p. 297 (1906).

1906–7

Abstract of a lecture on broncho-pneumonia in children. *St Bart's Hosp. J.*, **14**, pp. 88–9 (1906–7).

Abstract of a lecture on chorea. *St Bart's Hosp. J.*, **14**, pp. 88–9 (1906–7).

(A. E. Garrod and T. Shirley Hele.) A further note on the uniformity of H:N quotient in cases of alkaptonuria. *J. Physiol.*, **35**, *Proc. Physiol. Soc.*, pp. xv–xvi (1906–7).

1907

The initial stage of myositis ossificans progressiva. *St Bart's Hosp. Rep.*, **43**, pp. 43–9 (1907).

(A. E. Garrod and J. Wood Clarke.) A new case of alkaptonuria. *Biochem. J.*, **2**, pp. 217–20 (1907).

(A. E. Garrod and H. A. T. Fairbank.) A case of catarrhal appendicitis due to the presence of oxyuris vermicularis. *Lancet*, **ii**. p. 772 (1907).

1907–8

A lecture of chorea. Delivered at the Hospital for Sick Children, Great Ormond Street, W.C. *Clin. J.*, **31**, pp. 1–7 (1907–8).

A lecture in empyema in children. Delivered at St Bartholomew's Hospital. *Clin. J.*, **31**, pp. 193–7 (1907–8).

(A. E. Garrod and W. H. Hurtley.) On the supposed occurrence of uroleucic acid in the urine in some cases of alkaptonuria. *J. Physiol.*, **36**, pp. 136–41 (1907–8).

1908

Case of multiple rheumatic nodules in an adult. *Proc. roy. Soc. Med.*, **i**, Clin. Sect., pp. 13–14 (1908).

The Croonian lectures on inborn errors of metabolism. Delivered before the Royal College of Physicians of London on June 18th, 23rd, 25th and 30th, 1908. *Lancet*, **ii**, pp. 1–7, 73–9, 142–8, 214–20 (1908).

The initial stage of myositis ossificans progressiva. *St Bart's Hosp. Rep.*, **43**, pp. 43–9 (1908).

(F. T. Steward and A. E. Garrod.) Case of pyo-pericardium cured by drainage. *Proc. roy. Soc. Med.*, **I**, i, Clin. Sect., pp. 15–17 (1908).

1908–9

Critical review. The excretion in the urine of sugars other than glucose. *Quart. J. Med.*, **2**, pp. 438–54 (1908–9).

Individuality in its medical aspects. Extracts from sessional address before Abernethian Society. *St Bart's Hosp. Rep.*, **16**, pp. 18–21 (1908–9).

1909

Anomalies of urinary excretion (pp. 40–85); Uraemia (pp. 86–102). In *Modern medicine: its theory and practice*, Vol. 6, (ed. William Osler and Thomas McCrae). (1909).

Enterogenous cyanosis. In *A system of medicine*, Vol. 5, (ed. Sir Clifford Allbutt and Rolleston Humphry Davy), pp. 838–5 (1909).

Inborn errors of metabolism: the Croonian Lectures delivered before the Royal College of Physicians of London, in June, 1908. Frowde; Hodder and Stoughton, London (2nd ed.), (1909). Frowde; Hodder and Stoughton, London (1923).

Uraemia or meningitis? *Proc. roy. Soc. Med.*, **2**, i, Clin. Sect., pp. 169–75 (1909).

1909–10

Concerning intermittent hydrarthrosis. *Quart. J. Med.*, **3**, pp. 207–20 (1909–10).

Multiple peripheral neuritis in a child. *Proc. roy. Soc. Med.*, **3**, i, Sect. Stud. Dis. Child., pp. 38–40 (1910).

1911

A case of spondylitis deformans. *Proc. roy. Soc. Med.*, **4**, i, Clin. Sect., pp. 29–30 (1911).

On auscultation of the joints. *Proc. roy. Soc. Med.*, **4**, ii, Med. Sect., pp. 35–9 (1911); *Lancet*, **i**, pp. 213–14 (1911).

On the nature of the connexion between erythemata and lesions of joints. *Lancet*, **i**, pp. 1411–12 (1911).

Where chemistry and medicine meet. *Brit. med. J.*, **i**, pp. 1413–18 (1911).

1912

Lettsomian lectures on glycosuria. Delivered before the Medical Society of London. *Lancet*, **i**, pp. 483–8, 577–62, 629–33 (1912).

1912–13

The scientific spirit in medicine: inaugural sessional address to the Abernethian Society. *St Bart's Hosp. J.*, **20**, pp. 19–27 (1912–13).

(A. E. Garrod and W. H. Hurtley.) Congenital family steatorrhoea. *Quart. J. Med.*, **6**, pp. 242–58 (1912–13).

1913

Die diätetische Behandlung der Gicht. *Med. Klin.*, **9**, pp. 1153–8 (1913).

The dietetic treatment of gout. *Lancet*, **i**, pp. 1790–4 (1913).

Discussion on non-diabetic glycosuria. Opening paper. (Eighty-first annual meeting of the British Medical Association. Held in Brighton on July 23rd, 24th, and 25th. Section of Medicine.) *Brit. Med. J.*, **ii**, pp. 850–3 (1913).

(A. E. Garrod, Frederick E. Batten, and Hugh Thursfield) (ed.) *Diseases of children, by various authors*, Arnold, London, 1913. Garrod contributed the following sections: Disease as it affects children, pp. 1–4; Diseases of the ductless glands, pp. 560–84; Disorders of metabolism, pp. 585–602. Second edition, edited by Hugh Thursfield and Donald Paterson, 1929; Garrod's section on metabolic disorders was revised by E. A. Cockayne, pp. 530–57. Third edition, edited by Hugh Thursfield and Donald Paterson, 1934; Garrod wrote section on inborn errors of metabolism, pp. 583–92. Fourth edition, edited by Donald

Paterson and Alan Moncrieff, 2 vols., 1947–9. Fifth edition, edited by Alan Moncrieff and Philip Evans, 2 vols., 1953.

(Frew, R. S. and A. E. Garrod.) Glycosuria in tuberculous meningitis. *Lancet*, **i**, pp. 15–16 (1913).

1914

Address in medicine delivered at the eighty-second annual meeting of the British Medical Association. *Brit. med. J.*, **ii**, pp. 228–35 (1914).

Address in medicine: on medicine from the chemical standpoint. Delivered at the eighty-second annual meeting of the British Medical Association. *Lancet*, **ii**, pp. 281–9 (1914).

Clinical applications of pathological chemistry. *Trans. internat. Cong. Med., 1913*, Sub-sect. iii (a), Chem. Path., Pt. 2, pp. 71–81 (1914).

A discussion on the thymus gland in its clinical aspects. Opening paper.

(Eighty-second annual meeting of the British Medical Association. Held at Aberdeen on July 29th, 30th, and 31st. Section of Diseases of Children.) *Brit. med. J.*, **ii**, pp. 571–3 (1914).

The relations of chemistry to medicine. *Med. Press Circ.*, N.S. **98**, pp. 147–9, (1914).

(A. E. Garrod and Geoffrey Evans.) Sclerosis of the arch of the aorta, leading to obliteration of the pulses in the neck and upper limbs. *St Bart's Hosp. Rep.*, **50**, i, pp. 65–75 (1914).

1917

A variety of war heart which calls for treatment by complete rest. *Lancet*, **i**, pp. 985–6 (1917).

1919

Islands: a lecture delivered in the Aula Magna, Malta University, 21 Jan., 1919. Malta, Empire Press, for the University, 1919.

The laboratory and the ward. *Contributions to medical and biological research dedicated to Sir William Osler in honour of his seventieth birthday July 12, 1919, by his pupils and co-workers.* New York, Vol. I, pp. 59–69 (1919).

1919–20

A lesson in adaptation. *St Bart's Hosp. J.*, **27**, pp. 127–8 (1919–20).

1920

On learning medicine. *Guy's Hosp. Gaz.*, **34**, pp. 336–7 (1920).

The Schorstein Lecture on the diagnosis of disease of the pancreas. Delivered at the London Hospital Medical College on February 20th, 1920. *Brit. med. J.*, **i**, pp. 459–64 (1920).

1921

Children's diseases: a retrospect. *Arch. Pediat.*, **38**, pp. 129–40 (1921).

Sir William Osler, Bart., 1849–1919. *Proc. roy. Soc.*, Series B, 92, pp. xvii–xxiv (1921).

1921–2

(Leonard Mackey and A. E. Garrod). On congenital porphyrinuria, associated with hydroa aestivale and pink teeth. *Quart. J. Med.*, **15**, pp. 319–30 (1921–2)

1922

A bypath of medicine (congenital porphyrinuria). *Trans. roy. med.-chir. Soc., Glasgow*, **16**, p. 165 (1922).

In memoriam. John Wickham Legg. *St Bart's Hosp. Rep.*, **55**, pp. 1–6 (1922).

1923

Discussion on the modern treatment of diabetes. *Trans. med. Soc. Lond.*, **45**, pp. 3–4 (1923).

The Linacre Lecture entitled glimpses of the higher medicine. Delivered at Cambridge on May 5th, 1923. *Lancet*, **i**, pp. 1091–6 (1923).

1923–4

(A. E. Garrod and Geoffrey Evans.) Arthropathia psoriatica. *Quart. J. Med.*, **17**, pp. 171–8 (1923–4).

1924

The debt of science to medicine: being the Harveian Oration delivered before the Royal College of Physicians of London on St. Luke's Day 1924. Clarendon Press, Oxford, 1924.

The Harveian Oration on the debt of science to medicine. Delivered before the Royal College of Physicians of London on St. Luke's Day, October 18th. *Brit. med. J.*, **ii**, pp. 747–52 (1924).

Discussion on 'The aetiology and treatment of osteo-arthritis and rheumatoid arthritis'. *Proc. Roy. Soc. Med.*, **17**, 1–11, pp. 1–4 (1924).

1924–5

Examinations from the examiner's standpoint. *St Bart's Hosp. J.*, **32**, pp. 5–6 (1924–5).

1925

Alexander John Gaspard Marcet, Physician to Guy's Hospital, 1804–1819. *Guy's Hosp. Rep.*, **75**, pp. 373–87 (1925).

1925–6

(Leonard Mackey and A. E. Garrod.) A further contribution to the study of congenital porphyrinuria (haematoporphyria congenita). *Quart. J. Med.*, **19**, pp. 357–73 (1925–6).

1926

An address on the science of clinical medicine. Given at the Westminster Hospital, October 1st, 1926. *Brit. med. J.*, **ii**, pp. 621–4, (1926).

Science of clinical medicine. Delivered at the opening of the winter session of the Westminster Hospital on Oct. 1st, 1926. (Addresses to medical students. Abridged.) *Lancet*, **ii**, pp. 735–7 (1926).

1927

Congenital porphyrinuria. [Abstract.] *St Bart's Hosp. Rep.*, **60**, p. 186 (1927).

The Huxley Lecture on diathesis. Delivered at the Charing Cross Hospital, November 24th, 1927. *Brit. med. J.*, **ii**, pp. 967–71 (1927); Lancet, **ii**, pp. 1113–18 (1927).

1928

An address on the place of biochemistry in medicine. Delivered at the opening of the Courtauld Institute of Biochemistry at the Middlesex Hospital on June 14th. *Brit. med. J.*, **i**, pp. 1099–101 (1928).

The lessons of rare maladies. The Annual Oration delivered before the Medical Society of London on May 21st, 1928. *Lancet*, **i**, pp. 1055–60 (1928).

1928–9

J. F. Bullar. *(Corres.) St Bart's Hosp. J.*, **36**, p. 110 (1928–9).

1929

In memoriam. Sir Dyce Duckworth, Bart., M.D., 1840–1928. *St Bart's Hosp. Rep.*, **62**, pp. 18–30 (1929).

The power of personality. *Brit. med. J.*, **ii**, pp. 509–12 (1929).

1929–30

St Bartholomew's fifty years ago. Summer sessional delivered to the Abernethian Society, Thursday, June 5, 1930. *St Bart's Hosp. J.*, **37**, pp. 179–82, 200–4 (1929–30).

1931

The inborn factors in disease: an essay. Clarendon Press, Oxford, 1931.

1933–4

Chemistry and medicine. (Summary of lecture delivered to pre-clinical students.) *St Bart's Hosp. J.*, **41**, pp. 27–8 (1933–4).

1936

Congenital porphyrinuria: a postscript. *Quart. J. Med.*, **29**, pp. 473–80 (1936).

Index

AEG indicates Archibald Edward Garrod. Works by AEG and others appear directly under their titles.

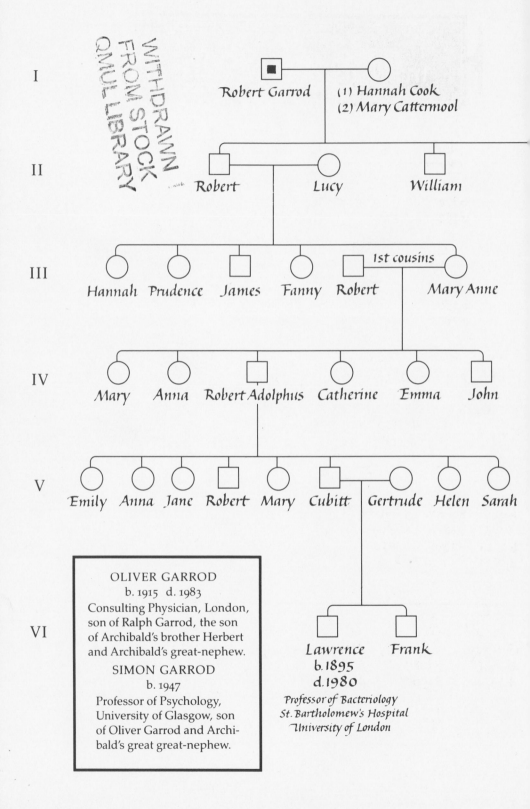

I   Robert Garrod   (1) Hannah Cook
                    (2) Mary Cattermool

II Robert   Lucy   William

III Hannah Prudence James Fanny Robert —1st cousins— Mary Anne

IV Mary Anna Robert Adolphus Catherine Emma John

V   Emily Anna Jane Robert Mary Cubitt Gertrude Helen Sarah

VI

OLIVER GARROD
b. 1915   d. 1983
Consulting Physician, London,
son of Ralph Garrod, the son
of Archibald's brother Herbert
and Archibald's great-nephew.

SIMON GARROD
b. 1947
Professor of Psychology,
University of Glasgow, son
of Oliver Garrod and Archibald's great great-nephew.

Lawrence   Frank
b. 1895
d. 1980

Professor of Bacteriology
St. Bartholomew's Hospital
University of London